INDEX

ABOUT THE AUTHOR

Sebastián Laza is an argentine economist, specialized in the interrelation between Cognitive Neuroscience and Behavioral Economics, with postgraduate courses on the subject at National Research University (Russia), Duke University (USA) and Copenhagen Business School (Denmark).

The aforementioned author is Professor and Executive Director of a Program in Applied Neurosciences to Management and Economics at the National University of Cuyo (Argentina).

Sebastián Laza is also the Coordinator of Neuroeconomics Area of Instituto Latinoamericano de Neurociencias Aplicadas (http://Neurosciences.online/), directed by the renowned Argentinean neuropsychologist PhD Roberto Bataller.

He has previously written *Neuroeconomics: The Disruptive Path* (Amazon KDP, 2018).

Additionally, he maintains the blog Neuroeconomía (http://seblaza.blogspot.com.ar/) with numerous articles on the fields of Neuroeconomics and Behavioral Economics, freely accessible to the general public.

THANKS

A warm thank to Nestor Braidot and Roberto Bataller, experts of international level in Neuroapplications to Business and Economics, who have given the motivational support to deciding to write a text referring to a completely new subject, and with very little standardized bibliography.

Also a strong thank for the remarkable Russian professor Vasily Klucharev, of the National Research University, academic director of an inspiring program on Neuroeconomics, which helped me to better structure all the topics of this book.

INTRODUCTION

Although it seems difficult to accept, economic decisions are practically taken before reaching consciousness, which is the place where economists always believed, was made the economic calculation, the rational cost-benefit equation. Today, innumerable neuroscientific papers on the subject all agree that the limbic (emotional) basis of economic decisions is very powerful, including the orbitofrontal cortex (which would emit the "value signal"), and always below the level of consciousness.

The neuroscientist John-Dylan Haynes (of the Max Planck Institute of Germany), perhaps one of the most advanced in these subjects, has shown, via neuroimaging, that we practically decided a choice seconds before we are aware of it, through refined subcortical and cortical mechanisms, which we do not control consciously. Haynes adds that there is evidence that a decision is encoded in circuits of the prefrontal and parietal cortex seconds before it enters consciousness, reflecting the functioning in the brain of high-level networks control areas, which begin to prepare a decision long before it enters the conscience, and that predetermines it in a very high percentage.

Undoubtedly, these ideas are quite revolutionary for Economics, since they go against the traditional consensus of the rational optimizing agent, besides since for decades it is known that our memory, the base of comparison to decide, works highly to unconscious level.

"The impression that we are able to freely choose between different possible courses of action is fundamental to our mental life. However, it has been suggested that this subjective experience of freedom is no more than an illusion and that our actions are initiated by unconscious mental processes long before we become aware of our intention to act". (Haynes, The Unconscious Determinants of Free Decisions in the Human Brain, Nature, 2008, pag. 1)

But these discoveries should not surprise us so much, it has been known for some time that the conscious part corresponds to only 10% of the energy consumed by the brain, while 90% is consumed by our unconscious, which confirms its enormous participation in decision-making, even if we do not realize it.

The brain consumes the same energy when it is doing research tasks that when it is sleeping. This situation is due to the fact that constantly, generally unconsciously, it is receiving and processing internal and external information. It is always alert, constantly evaluating alternatives to possible future situations that may pose a danger to the survival of its owner.

We do not know why we choose what we buy

Going to the bottom of the subject, the consumer does not know at all why he chooses what he buys. The decision is made, in large part, below the threshold of consciousness, where our most instinctive biology and our most emotional parts carve strong. In the unconscious, the interest for the product, the purchase intention and the loyalty to the brand are elucidated. These elements correspond to the construction, partly induced by the promotional campaigns, of desires and loyalty.

The mechanism works in the following way: certain sensory stimuli (induced by large corporations) activate deep areas of our brain. The reward system (limbic and subconscious), especially the nucleus accumbens, is put into action and drives to seek food, sex and safety, the three basic pillars of human survival. The brands, which I repeat are creations of large corporations to induce, seek to activate brain areas that regulate the sense of belonging, making us part of a group, a tribe, a community. All this, together with the natural tendency to imitate and / or empathize with everything that surrounds us (mirror neurons), leads us to consume much less rationally than we believe, pulling down the dogma of free choice, the sovereignty of the consumer.

Our rationality has a very high unconscious component, although it is difficult to accept it, especially those people who think they are too rational and self-controlled. The day that the vast community of social scientists fully understands these concepts, which come from Cognitive Neuroscience, they will have to start rewriting many chapters of university books. And believe us that time is getting closer.

Bounded Rationality

Additionally to the unconscious side of the decision making, the human being makes decisions in a context of limited rationality, that is, subject to a huge amount of biases and sub-information that lead him to behave, in many cases, in a suboptimal way from the point of view of what Neoclassical Economics prescribes. Behavioral Economics has been showing this phenomenon clearly for at least four decades, with

Nobel Prizes Simon, Kahneman and Thaler as main banners.

However, in recent years, the disruptive confluence of Neuroscience, Psychology and Economics has built a hybrid field called Neuroeconomics, which with methods different from the traditional ones has added even more evidence about biases and limited rationality: human decisions related to consumption, investment, saving, among many others, are not based exclusively on the calculating and maximizing ratio of neoclassical homo economics (pre-frontal cortex), but rather more uncontrollable and automatic elements often come into play, such as emotions, feelings, intuitions and biases (limbic system).

Kahneman, winner of the Nobel Prize in Economics in 2002, describes: *"the most important characteristic of a human being is not that he reasons poorly, but often acts instinctively; and the behavior is not guided by the calculations that can be made, but by what is seen at the moment when the decision has to be made"*.[1]

But it happens that these human "fragilities" (the departure from full rationality when making economic decisions), so frequent in our daily behavior, until very recently were very little taken into account at the time of analyzing and modeling economic processes. For example, Vernon L. Smith, also Nobel Prize in Economics in 2002, states that until recently, economics was considered a non-experimental science that had to rely on observation of the real world more than in laboratory experiments, and cites the following words of the renowned Paul Samuelson: *"Due to the complexity of human and social behavior we cannot hope to have the precision of the physical sciences, chemists and biologists. As astronomers, we should content ourselves with observing."*

However, and luckily, we now know that this appreciation of Samuelson was wrong, and over the years the evolution of neuroscientific technology has allowed us to overcome limits that were believed impossible to achieve.

In this way, during the last fifteen years, neuroscientists and "open minded" economists around the world are focusing their research precisely on the interaction between the rational and the emotional brain, when people make decisions related to scarcity and money, giving rise to a new field of study called Neuroeconomics. Let's look at some recent definitions of this hybrid term:

Says Paul Zak[2]:

Neuroeconomics is an emerging transdisciplinary field that uses neuroscientific techniques to identify the neural substrate associated with economic decisions. "Economic" should be interpreted here in the broadest sense, as any decisional process (human or non-human) that is given by the evaluation of alternatives.

According to Paul W. Glimcher and Aldo Rustichini[3]:

Neuroeconomics has tried to put together the theory and methodology of diverse areas such as economics, psychology, neurology, cognitive science, cognitive neurology, mathematics, statistics, behavioral finance and decision theory to create a model of human behavior that not only explains, but also predicts how people make decisions.

According to Kevin McCabe[4]:

Neuroeconomics is an interdisciplinary research program whose goal is to build a biological model of decision making in economic environments. Neuroeconomists are asking how the embodied brain allows the mind (or group of minds) to make economic decisions.

Words more, words less, the above definitions explain that by Neuroeconomics we mean all those efforts to try to elucidate the true mechanisms that underlie our economic decision making, which until now economists have almost always been very rational. And to this new way of investigating economics, majority of scientific community already grants it the character of a "respectable research program".

Refining the concept a bit more, today we talk about two different Neuroeconomics, or rather, two different research programs within the same field, which we should already be clear from the beginning of this work:

- Behavioral Economics in the Scanner (BES)
- Neurocellullar Economics (NE)

In this book we will analyze both branches, although a little more the BES branch, the most advanced in terms of field research, which is the version of Neuroeconomics that tries to empirically test the concepts of Behavioral Economics. The other branch, the NE, opens a bit of Behavioral Economics to understand the mechanisms by which the brain comparatively evaluates possible alternatives with scarce resources, and that, under the guidance of Paul Glimcher, is increasingly gathering scientific respect.

Neuroeconomics has aroused great expectations since its inception. For example, let's mention a reflection by

Colin Camerer, when this research program started (year 2000):

"Something tells me that within ten years, the entire digital universe is going to seem like pretty mundane stuff compared to the new technology that right now is but a mere glow radiating from a tiny number of American and Cuban (yes, Cuban) hospitals and laboratories. It is called <u>brain imaging</u> and anyone who cares to get up early and catch a truly blinding twenty-first-century dawn will want to keep an eye on it...If I were a college student today, I don't think I could resist going into neuroscience. Here we have the two most fascinating riddles of the twenty-first century: the riddle of the human mind and the riddle of what happen to the human mind when it comes to know itself absolutely."[5]

The ten years have passed and, realistically, still the important advances in Neurosciences have not manifested markedly in Social Sciences, particularly economics; remember that traditional (highly rationalist) thinking is still strong in our science. Additionally, neuroimaging (brain imaging), as an econometric research instrument, has still some problems that need to be overcome, which do not invalidate it, but which do require the complement of other field instruments, such as Transcranial Magnetic Stimulation (TMS), and other techniques that we will also mention later. However, the seeds for change are already scattered, and in this Neurosciences have been determining.

What is clear is that today, thanks to all these advances, the human brain can be analyzed by making decisions in real time: purchase, investment, negotiation, etc., and with a high degree of detail. And

12

you do not need to be a futurologist to predict that, sooner or later, economists should at least become familiar with these issues, as is already happening in Management with Neuromarketing,

Neuromanagement and Neuroleadership. That is to say, the change will come, that is inevitable. What nobody knows yet is how deep it will be within economic science, it will depend a lot on the methodological subject in economics (epistemological), of the dominant and emerging research programs, a topic to which we will dedicate an entire chapter in this book.

In another fact that is not minor, today the vast majority of economists are aware of the high degree of mathematization that our science has acquired. Studying economics - both at the undergraduate and graduate levels - has become increasingly complex. And although several are proud and fully in agreement with this course, many others note the abuse that has been committed in recent years with the excessive mathematization of economic theory, mathematization that hides behind, underlying - as we economists like to say - the strong neoclassical rationalist thought that still survives within economic theory, which seems not even have known that modernity and positivism / rationalism, as paradigms in Social Sciences, were long ago replaced by postmodernity, the complex systemic and so many innovations that could have improved it.

That is, today many economists are wondering if it was really worth such a degree of theoretical mathematization to understand the economic functioning of the world. And this luck of boredom, also plays in favor of many economists begin to "look with affection" to Neurosciences and the possibility of analyzing and modeling human beings making economic decisions, via methods other than hypothetical-deductive, so rationally logical and so based on mathematics and its abstractions, that it was always quite remote from real man, flesh and blood.

Today Economics can use the contributions of Neurosciences to access the black box (the mind) of economic agents -consumers, investors, savers, etc.-, instead of relying so much on the equations and their ultra-maximizing assumptions. It is not a small thing of what we are talking about.

But traditional thinking, ultra-rationalist and maximizing, has been and is powerful and influential in economics and will be difficult to qualify, let alone dethrone. Most economists have heard from their professors that our science is converging more and more towards "the micro fundamentals of macroeconomics", meaning that it was not going to be seen with seriousness that theoretical economist who raised some macro postulate without putting together his theoretical model from the micro, obviously the micro of the neoclassical homo economicus, the one that Neurosciences today discredit.

Let us quote the following criticism of utilitarianism (the cornerstone of neoclassical economics) by the brilliant historian of economic analysis J. Schumpeter:

> *"The psychology actually used [...] was always individual psychology, introspective, and the most primitive type, rarely endowed - if it ever was - of more than a few simple hypotheses about the reactions of the individual psyche. This procedure was called empirical [...]. There was nothing*

14

"experimental" or inductive, and in reality it was not very realistic, despite the programmatic statements, the war cries and the invocations of Francis Bacon."[6]

However, the discourse of high rationality in terms of economic theory, even in our days, has many adherents, most of them framed in the so-called classic / neoclassical / new classics schools and all its branches and sub-branches, both micro level as macro.

But these attacks of Neuroeconomics and Behavioral Economics, although they are the strongest, are not the first against the ultra-maximizing rationality. Throughout the history of economic analysis there have been high-sounding voices to the dominant discourse, mainly from the so-called Keynesian School, where both JM Keynes and his followers emphasized that economics does not always walk along paths of high rationality and that for example, sometimes there are situations of collective panic, which imply paralysis of investment and consumption even under conditions that should allow rational men to return to a situation of equilibrium, that is to come out of a crisis.

In short, behind the Keynesian thought and its ramifications lies the idea that individuals and companies are not 100 percent maximizing, for different and debatable reasons, but not always maximizing. And therefore, the Keynesians have always tried to model human economic behavior by contemplating these "presumably irrational" avatars, unlike mainstream thinking.

And while Keynesianism has been powerful in its critical potential - and also in its ability to influence politics - it has been weak at the theoretical level;

15

although it is necessary to recognize that, specifically in the subject rationality, it did not count on the weapons that today Cognitive Neuroscience offer to base their ideas.

But beyond this kind of "unfulfilled promise" of Keynesianism in the task of "thoroughly" modifying the current theory, today the hopes of improvement have passed to Behavioral Economics and especially to Neuroeconomics, particularly when more and more economists, worldwide, begin to familiarize themselves with the advances that come from the Cognitive Neuroscience.

For example, Nobel Prize Simon has been key in this crusade with his theory of "bounded rationality" back in the '70s, which were the basis of what we now call Economics of Behavior.

And since Neuroeconomics, today it is clearly shown that the decision that drives a purchase is not a completely rational process, but in most cases is relatively automatic and derived from metaconscious forces, that is, people incorporate many more things to the purchase decision that the simple cost-benefit analysis that we use in micro-and now also in macroeconomics-, issues beyond the reason that obviously are not calculated and more than obviously cannot be mathematized in the traditional way.

And to illustrate a little more, we are going to mention other ideas-force that come from Neuromarketing, which are perfectly applicable in Economics, especially in everything related to the consumption function at the macro level or to the theory of demand at the micro level[7]:

"According to scientists, the brain areas of rationality cannot function isolated from the areas of biological-emotional regulation. The two systems communicate and affect the behavior jointly, and consequently, the behavior of the people ".

"Moreover, the emotional system (the oldest area of the brain) is the first force that acts on mental processes, therefore determines the direction of decisions."

"The fragrance of a perfume, for example, can evoke different sensations. If the client associates it with painful experiences or with a person with whom he does not sympathize, it is very likely that he will not buy it, even when the price-quality-brand ratio is reasonable ".

"These and other associations, like most mental processes, are verified in the metaconscious plane and force us to find new tools that allow us to access that disordered set of emotions, memories, thoughts and perceptions that determine the decisions of purchase and consumption, and that most of the times the client does not know ".

These ideas, coming from Neuromarketing, are just a sample of how far apart human beings are from full rationality in consumer decision-making, and as a consequence of that, also from the economic-maximizing models of functions. The creative question of Neuroeconomics is going to be given by the way in which it incorporates the "non-rational" variables into the models, which allow modeling "more human" consumers. The challenge is great.

And to mention one more example of the potential contributions of Neuroeconomics to traditional theory, let's take the case of Game Theory, very fashionable in recent times. In a paper written by the neuroeconomists Camerer, Loewenstein and Prelec[8], it is argued that economic theory has assumed (without implying an interest in Neuroscience) that agents possess a ToM module (Theory of Mind), and that they can "to mentalize" (deduct from the actions of other agents, and quite correctly, what their preferences and beliefs are). However, since the modern advances in Neuroscience, there is accumulated evidence that "mentalizing" is a specialized skill and modularized in specific brain regions, and that this ability effectively exists, but in varying degrees of person to person, while game's theory generalizes for all type of economic actor.

Game theory is based on the assumption that people are capable of predicting the actions of others. The most fundamental concepts in that field - Nash equilibrium, late induction, and iterated elimination of dominated strategies - are based on this assumption. These concepts require that people be able to see the game from the perspectives of other players, for example, that they understand the motives and beliefs of others. However, traditional economists (unlike neuroeconomists) know little or nothing about what allows people to "put themselves in the shoes of others" and about how this ability interacts with their own preferences and beliefs. In fact, experimental evidence suggests that many people do not obey the traditional concepts of game theory and often behave as if they - counter-objectively - believe that others are going to play with dominant strategies.

In this way, today Social Neuroscience and particularly Neuroeconomics provide new ideas about the neural mechanism that underlies our ability to represent intentions of others, beliefs, and desires, and also about the ability to share the feelings of others, called "empathy"; it is then from these new approaches where game's theory should rethink their assumptions.

At this point, we believe the purpose of this work is clear: to illustrate some of the main advances that Cognitive Neuroscience have been showing towards the decision-making process - mainly what has been discovered about our emotional brain (limbic system) and its complex interaction with our most rational areas (prefrontal cortex) - and, what is even more important, also illustrate the enormous possibilities that are opening up today in Economics to find an answer to many of the anomalies existing in the traditional theory, giving richness to the debate on possible changes in the dominant research program, although this hypothesis may seem too ambitious for the current achievements of Neuroeconomics, it does not take away future possibilities.

REASON, EMOTION & ECONOMIC THEORY: A RETROSPECTIVE VIEW

The assumption about the high rationality of economic agents has been key to the construction of modern economic theory, which began to take shape, as a separate science, approximately with the neoclassical (Jevons, Walras, etc.) during the nineteenth century. In terms of Lakatos, one of the most influential epistemologists of the twentieth century, all science has a hard core, which is very difficult to refute, to modify, and in which there are certain premises that nobody usually discusses, and all accept them as basal foundation from where the current models start. And the premise of rationality that prevailed in economics is that of the hyper-maximizing human being, always tending towards quasi-perfect cost-benefit evaluations as the basis of each economic decision; this is perhaps the fundamental assumption on which the neoclassical built modern economic theory, and which is still valid today, beyond the numerous criticisms received over the past two centuries, with the School of Behavioral Economics and Neuroeconomics among the most recent critics.

Throughout this chapter, we will try to walk the evolutionary path that has transited the concept of rationality in economic theory, emphasizing some of its main critics, from JM Keynes to the Nobel prizes Simon, Thaler and Kahneman, and to modern neuroeconomists like Glimcher, Camerer, Zak, etc., to name but a few of the most important critics.

1.   Adam Smith: Reason vs Passion

The first economists began their task when psychology still did not exist, which is why they acted in some way as psychologists. Hume's work [9] is largely devoted to analyzing human knowledge from a perspective that we would consider today as a field of psychology, and it is not precisely a simplified and monolithic vision that serves as a support for the neoclassical model, but rather, applying introspection, describes a much more complex and real human being.

In this line of thought is the work of Adam Smith "Theory of Moral Sentiments"[10], which is a detailed (if basic) analysis of human psychology. Following the Platonic distinction, Adam Smith differentiates two systems in the human being, one affective, linked to the passions and the most primitive feelings, and another superior, which controls, in the manner of an impartial spectator, the first:

When I strive to examine my own behavior, when I endeavor to pronounce judgments on it, either to approve it or to condemn it, it is evident that in such cases it is as if I were divided into two different persons, and that I, the examiner and the judge, I embody a man different from the other me, the person whose conduct is examined and judged.
The first is the spectator ... The second is the agent, the person that I designate as myself, and from whose behavior I tried to form a feeling, as if it were a spectator's. The first is the judge, the second the person who is judged...

When we are about to act, the avidity of passion will rarely allow us to consider what we do with the dispassion of an intelligent person...

In the "Wealth of Nations", according to Nobel Prize winner Simon[11]:

"...the rationality that Smith describes is that of common sense every day. This follows from the idea that people have reasons to do what they do, and that this does not depend on an elaborate calculation of utility."

But not all the outstanding classics thought the same way as Smith, at least not in regard to how to model human beings when trying to do economic theory. For example, let's take Stuart Mill and his concept of homo economicus[12]. The main ideas in this regard are the following: first, Mill recognizes that there is a part of human behavior where obtaining wealth is not the main objective. Now, there are other departments of human affairs where the acquisition of wealth is the main purpose; Economics deals with this second category, so that it abstracts from all human passions and motives except the desire for wealth and the aversion to work. The man thus described is a fictitious man, and Mill himself is aware that the economic sphere is only a part of human behavior. However, he recommends that economics proceed to abstract and work with this fictitious man, who seeks to obtain "the greatest possible amount of wealth with the minimum possible work and self-denial".

And in general, it is pertinent to note that along with Stuart Mill, two other classic theorists, also important at the time, such as Senior and Cairnes, coincide in the search for maximum wealth with the least possible effort as one of the driving principles of the man. The coincidence is not accidental, but responds to the influence in the England of s. XIX exercised the philosophical current of utilitarianism.

In this way, classical economists seem to have no unanimity about how human rationality should be taken when doing economics, on the one hand there were, among others, David Hume and Adam Smith, the latter called the "father of Economics", who introduced the principle of personal interest, but with the above-mentioned limitations (especially in his "Theory of the Moral Sentiments"), but on the other hand there were Mill, Senior, Cairnes, among others, closer to the utilitarian currents that were going to impact fully in the subsequent school, the neoclassical ones.

2. The Animal Spirits of J.M.Keynes

John Maynard Keynes also departs from the concept of rationality when he asks how it can be that even when the rational analysis of investment projects shows its inconvenience, economic agents decide to invest despite the high probability that the project will not turn out to be profitable and that can bankrupt the investor. It supposes that this is due to the "animal spirits", which are something like waves of optimism and pessimism that envelop society alternately and that move us to action for the pleasure of doing things, beyond what the cold cost-benefit calculation says. In addition, the inflexibility of falling wages, the monetary illusion, the inability of businessmen to adequately formulate their expectations and the trap of liquidity - all Keynesian concepts - are manifestations of the withdrawal of full rationality on the part of the economic agents, who make economics diverge naturally from full employment and public policies that restore it are necessary.

The contribution of Keynes to economic science is very important, basically because of the degree of influence he had and still has today in applied macroeconomic policy, especially in the short term. And of course, it helped to introduce into current economic theory certain aspects that make the true rationality of man, not the ad-hoc that Robbins enthroned, and that comes from the utilitarians. That is why we are going to do a more detailed analysis of this economist.

To begin with, it is said that Macroeconomics was born as something separate with Keynes, that is, it begins to differentiate the micro from the macro. During the nineteenth century and the first decades of

24

the twentieth century the vast majority of neoclassical economists - Jevons, Walras and Menger, and their disciples Marshall, Edgeworth and Pareto - focused mainly on the study of microeconomic issues, although it is true that some of they were also interested in topics of a macro nature. With respect to the aggregate functioning of economics, there was a certain consensus regarding some basic principles, among which the validity of the Quantitative Theory of Money - in its Marshallian version, for example - was the validity of the price and wage flexibility guaranteed by the full employment and the effectiveness of Say's Law.

But in 1936 John Maynard Keynes published The General Theory of Employment, Interest and Money[13], one of the most influential economics books of the 20th century. The appearance of Keynes's book was of crucial importance due to two reasons. In the first place, this work supposes the birth of the Macroeconomics in its current form where Keynes - and from it, later the Keynesian economists - elaborates macroeconomic models proper, characterized by a particular way of adding markets, goods and economics agents. Second, the subsequent dissemination of the ideas contained in the General Theory by authors such as Samuelson and Hicks broke the existing agreement on macroeconomic issues referred to above (the flexibility of prices and wages, Say's law, etc.).

Two types of factors can be distinguished that contribute to the development of Keynesian thought: on the one hand, the high unemployment rates in England and the United States in the 1930s, which led economists to question the causes and remedies of this pathology. Second, Marshallian microeconomics was also being questioned by economists such as Joan Robinson, Chamberlin, Kahn, and Harrod. In short, John Maynard Keynes knew how to elaborate the theoretical framework that supported and justified, in a more or less coherent way, two beliefs that were accepted by economists and that classical economics of orthodox and hyper-rationalist tendency was not able to adequately explain:

- on the one hand, that the observed unemployment was involuntary;
- on the other, that fluctuations in aggregate demand had a strong impact on income and employment.

In particular, the General Theory linked both ideas and offered a plausible diagnosis and remedy of mass unemployment: the cause of unemployment was the insufficiency of effective demand. The solution, on the other hand, was in the stimulation of the latter. Keynes supports his analytical construction on principles radically opposed to those that maintain the classics, a term with which Keynes designates, disdainfully, all those who accept the basic premises on money, prices, wages and Say's Law detailed above.

The alternative principles on which Keynes works are the following: first, he does not accept the Quantitative Theory of Money because the demand for money is not directly related to rent (for the reason transaction) but also, inversely, to the type of interest (Keynes - great speculator in the stock market - highlights the speculation motive to demand money); secondly, it postulates that there are certain rigidities in prices and wages, and in particular that the nominal salary is

26

rigid due to institutional aspects such as the unions or the monetary illusion of the workers; and, finally, defends the invalidity of Say's Law since it is the demand that creates its own offer and not the other way round (or, in other words, nothing guarantees that the saving equals the investment at the level of full employment) .

The conjunction of these premises gives rise to one of the crucial implications of the General Theory: economics can be placed for long periods of time in a situation of equilibrium with unemployment (that is, the most irrational that can be for the classics); given that nominal wages are rigid and that Say's Law is a fallacy, economics alone will not return to the level of full employment. Therefore, the active intervention of economic policy becomes necessary. However, Keynes doubts the effectiveness of monetary policy given that, in his conceptual apparatus, investment is rigid and the demand for money -at low interest rate levels- is quite elastic with respect to the interest rate, which is why the prescription of economic policy is also immediate: the impulse of the aggregate demand must be carried out by means of an expansive fiscal policy (and, therefore, opposed to the orthodox dogma of the balanced budget).

At this point we are already appreciating the way in which Keynes begins to refute the dominant hyper-rationalist / maximizing economic theory until now:

- first, using macro variables instead of micro variables;
- second, from an acute and controversial reasoning, obviously introspective - there was no neuroimaging or EMT - about how human

beings, especially businessmen and consumers, make certain decisions: prices and wages are rigid in the short term, the expansive monetary policy in extreme situations has no effect on the expectations of economic agents, entrepreneurs sometimes invest without necessarily looking at profitability in the short term; in short, a whole series of aspects that Keynes observed happened in economics (and that the traditional theory did not contemplate), and that when beginning to be debated, and inserted in the theoretical models, they began to bring a little closer to the man of economic theory the man of flesh and bone, the real human being, not Robbins.

The publication of the General Theory, and the certain air of ambiguity with which it was written, generated an enormous volume of works that tried to unravel the authentic message of Keynes. The work of Patinkin (1956) [14] , which analyzes both Keynesian and Neoclassical thought in detail and depth, must be highlighted, so that, on the one hand, it provides a clear exposition of Keynes' theory; on the other hand, it shows the logical coherence of neoclassical propositions. In any case, and as we have already said, the influence of the Keynesian contribution was immense, both in the academic field and in that of economic policy. Certainly, most economists, during the 1950s and 1960s, developed their contributions within the framework of Keynesian thought, theoretically refining or empirically contrasting some of their propositions. In the applied field, the ideas of Keynes - and in particular the prominence attributed to fiscal policy - constituted the new orthodoxy that replaced the traditional one in most of the Western countries.

The interpretation of Keynes's thought that can be considered dominant is the so-called neoclassical synthesis of Hicks and Modigliani, popularized in its graphic version by the IS-LM curves. The model accurately captured the central message of the Keynesian contribution: the fact that prices and wages adapt slowly (that is, irrationally for the classics) to the mismatches between supply and demand. On the other hand, the qualification of neoclassical was due to the fact that the economic environment was perfectly Walrasian: markets were competitive; there were no externalities or imperfections in the information available to agents. The IS-LM model soon achieved great success: in fact, it has exercised an undeniable influence on the profession and has been incorporated into the vast majority of Macroeconomics textbooks for its (apparent) simplicity, elegance and versatility; it also continues to be used in recent manuals. The model suffers, however, from certain limitations that hinder its understanding and generate confusion in those who study it in depth, as is its timeless nature since it is a model of comparative statics and, therefore, not explicitly dynamic, and also its omission of the role of expectations. In addition, it is surprising that it is a Walrasian general equilibrium model in which there are rigid prices and salaries, at least in the short term.

But beyond the limitations mentioned, it is undeniable, from Keynes, the advance of economic theory to consider in their models much more realistic assumptions about how consumers, investors, and economic actors in general reason and make their decisions, against the excessive oversimplification of the Jevons, Marshall, Robbins and all those who, for intellectual and scientific convenience, assumed machine-men at the time of building the theoretical models of economic decision-making.

3. Simon's Bounded Rationality

Simon gives account of his critics to the principle of rationality in the decisions of the entrepreneurs, from a series of works that made him the winner of the Nobel Prize[15]. Define his idea of "bounded rationality" in the following terms:

The task, then, was to replace the classical model with one that describes how decisions could be made (and probably actually were) when the alternatives of search had to be sought out, the consequences of choosing particular alternatives were very imperfectly known both because of limited computational power and because of uncertainty in the external world, and the decision maker did not possess a general and consistent utility function for comparing heterogeneous alternatives.

Several procedures of rather applicability and wide use have been discovered that transform intractable decisions into tractable ones. One procedure already mentioned is to look for satisfactory choices instead of optimal ones. Another is to replace abstract, global goals with tangible subgoals, whose achievement can be observed and measured. A third is to divide up the decision -making tasks among many specialists, coordinating their work by means of a structure of communications and authority relations. All of these, and others, fit the general rubric of "bounded rationality".

Simon then opens a gate for the reformulation of the firm's theory and business decisions, which attempts to modify the neoclassical model. Instead of optimizing in the way that neoclassical theory assumes, economic agents set a goal. When they achieve it, even if it is not optimal, they feel satisfied with it and do not seek to optimize. The men of flesh and bone have limited capacities to acquire knowledge and to make calculations, and to predict the behavior would require the participation of psychologists and sociologists, in addition to the economists.

And in line with Simon's concept of limited rationality, we have Akerlof with his concept of cognitive dissonance, which also illustrates us about behaviors contrary to the supposed individual rationality that governs economic decision making, for example in situations where, those who make decisions, do not know their preferences well, or are too influenced when they act as part of closed groups to external points of view. The due obedience, which leads someone to do things that displease him for pleasing the superior, is an extreme example.

However, it must be recognized that, after all that has been said in previous pages about the enormous degree of current penetration of micro fundamentals and rational expectations in macroeconomics, and even though Simon has won a Nobel prize in economics, it would seem that the concept of limited rationality, in the '70 and '80, could not succeed in changing the course of traditional modeling, and therefore of the dominant paradigm at that time. But it really was a valuable attempt by Simon, and then continued by the other Nobel Prizes Kahneman and

Thaler, with great success in the '90, with the Behavioral Economics, and now with the Neuroeconomics.

4. Kahneman, Thaler and the Behavioral Economics

The revival of Psychology within Economics is translated into the current of thought that is mainly covered under the denomination of Behavioral Economics, which is disseminated and generalized with the awarding of the Nobel Prize in Economics in 2002 to Kahneman, who receives it in conjunction with Vernon Smith, whose branch, although related to the previous one, is called Experimental Economics. Both notables theoretical define two types of cognitive processes: System 1, which they call intuition and System 2, reasoning:

"The operations of System 1 are fast, automatic, effortless, associative, and often emotionally charged; they are also governed by habit, and are therefore difficult to control or modify. The operations of System 2 are slower, serial, effortful, and deliberatively controlled: they are also relatively flexible and potentially rule-governed."

"Utility cannot be divorced from emotion, and emotions are triggered by changes. A theory of choice that complete ignores feelings such as pain of losses and the regret of mistakes is not only descriptively unrealistic, it also leads to prescriptions that do not maximize the utility of outcomes as they are actually experienced.[16]"

To be rigorous, and beyond their great coincidences, the substantial difference between Behavioral Economics and Experimental Economics is that the first is based on the assumption that incorporating psychological principles will improve economic analysis, while the second presupposes that incorporating methods of psychology (for example controlled experiments) will only improve the testing of economic theory. Then we will devote a few paragraphs mainly to Behavioral Economics, which is, of the two branches, the one that has had the most impact at the theoretical contributions level in Economics.

In a landmark book on Behavioral Economics, Camerer and Loewenstein (2004) [17] summarize the main findings of this current. The method used by economists and psychologists working in the aforementioned line is mainly the active experiment, that is to say the one that is carried out on a group of chosen people, to which they are subjected to questions related to the subject under study, it is repeatable and it can be analyzed statistically, although the other methods used by Economics in general are also used. However, what distinguishes this current is the use of knowledge that comes from psychology to analyze economic behavior.

In a very interesting work, the Peruvian economist Ernesto López [18] points out some of the current conceptual contributions of Behavioral Economics. Following Mullainathan and Thaler (2000)[19], he asserts that behaviorists criticize the neoclassical economic paradigm, since it is based, in terms of its assumptions about agents, on three attributes, at least highly debatable:

- unlimited rationality;
- unlimited will;
- unlimited selfishness.

With regard to the attribute of unlimited rationality, and making a bit of history, it is necessary that, as early as 1955, Herbert Simon, whom we mentioned in the previous section, criticized the economic models that adopted the assumption of agents with unlimited capacities for processing information, which led him to coin the term bounded rationality to describe a more realistic view of human capacity for information processing.

We have already stressed that, according to this vision, human beings face restrictions of mental capacity and time and, therefore, will not always be able to solve complex problems optimally. Consequently, a "rational" strategy against these restrictions may be the adoption of practical rules that allow people to economize in the use of time or their mental faculties. But, just as this rational strategy can facilitate complex decisions, it can also lead to systematic errors, that is, repeated ones, as shown by Kahneman and Tversky[20].

Deviations from the assumption of rationality can occur with respect to judgments -based on beliefs- of the agents, which leads to situations of overconfidence, anchoring, extrapolation and judgments about the probability of future events based on limited but available information. They can also occur with respect to agent options, described by the prospect theory of Kahneman and Tversky. Two important concepts in this theory are those of "aversion to losses" and "mental accounting". The concept of "aversion to losses" suggests that people are more sensitive to decreases in their well-being than to increases in it, or in other words, it has been empirically verified that, in many cases, the decrease in utility associated with a loss is greater than the increase in utility associated with an equivalent gain.

For its part, the concept of "mental accounting" coined by Thaler[21], refers to situations in which agents, in the face of repetitive events with uncertain results, treat them as independent results and adopt a strategy for each of them, instead of to consider them as a single pool of events and adopt a general strategy. An example of mental accounting, collected by Camerer[22], is the behavior of taxi drivers in New York City. As in many other countries, many New York taxi drivers pay a fixed rent for the use of a taxi, and keep the rest of the income they earn. In this situation, the "rational" strategy of optimization would be to work more during the days of high demand (days with bad weather, or days when there is a big event in the city) and slightly less during days of low demand.

However, if the drivers evaluated each day independently, and compared the income of the day with a pre-established standard, they could end up working more hours, precisely in the days of low demand, something quite unsound from the neoclassical theory, but which is precisely the most usual empirical finding.

In relation to the second attribute, that of unlimited will, there are numerous examples of situations in which it can be affirmed that agents effectively know what is best for them, but do not opt accordingly due to problems of self-control. These deviations occur in the case of addictions, but also in usually less severe cases, such as bad eating habits, sedentary lifestyle or simple procrastination (leave for tomorrow what can be done today), something that usually happens to the majority of people.

Finally, the attribute of unlimited selfishness is also rebuttable and, happily, innumerable examples of altruistic behavior can be found, including the relative success of many national collections and volunteerism in charities.

Undoubtedly, it is quite clear, after all these examples, that behaviorists reason and theorize following a line of argument very similar to Simon's ("bounded rationality"), and obviously in tune with Neuroeconomics, but with the difference that their models were born based on Psychology more than Cognitive Neuroscience, unlike Neuroeconomics, which has stronger solid science foundations. However, behavioral contributions have been growing, and with a high degree of acceptance of mainstreams (two Nobel prizes), especially today that their theoretical developments are being provided with foundations in Cognitive Neurosciences, which gives them more rigor.

5. Neuroeconomics

While the conception of the neoclassical model starts from the idea that human beings have well-defined objectives that they try to obtain, the first findings of Neuroeconomics confirm the idea that in a person there are at least two decision centers, one from the "deliberative" system, located in the cerebral cortex, and another "affective" system, located in the inner part of the brain, that is, in its limbic part; and both systems interact permanently.

We return this way to the beginning, when Adam Smith (from introspection, not from Neurosciences) spoke of a confrontation between our passions and what he calls "impartial spectator" (Smith used a well-directed psychology, but very rudimentary). Although the neoclassical model starts from the premise that consumers optimize their utility and entrepreneurs maximize their profits, in a scenario of perfect information, this has not been the case at the beginning of economic science on the one hand (Adam Smith and others classics), and on the other there have been divergent opinions with that model for a long time (the aforementioned Hutchison, J.M.Keynes, Simon, among others mentioned throughout this chapter).

However, it is generally recognized that the neoclassical model has worked reasonably well, although we believe it must be discussed again in its fundamental premise (quasi-perfect optimizing rationality) in order to build a better economic theory. Moreover, we could say that any human decision that is theoretically modeled should be proposed in such a way as to maximize together the rational and the emotional that coexist in the human being, in order to reach a real and complete balance; where an alternative, albeit with criticism, could be the following model, by Loewestein and O'Donoghue[23].

In their work, both economists raise both the deliberative (rational, system 2) and affective (emotional, system 1) systems, since both underlie human behavior, and assume that the human being faces a function to minimize, which is the cost of their behavior. One part of the cost is the difference between what the deliberative system wants and what it ultimately obtains and another part of the cost is the effort that the deliberative system (led by the dorsolateral prefrontal cortex) must make to spur the impulse to act certain way (that comes from the affective system).

$$[U(x^D, c(s), a(s)) - U(x^A, c(s), a(s))] + h(W,\sigma)[M(a^A, a(s) - M(x, a(s))]$$

where U is a utility function, x the chosen course of action, of a set X, the supra-indexes D and A indicate the optimal behaviors for the deliberative and affective systems respectively, s is a vector of stimuli, a(s) and c(s) are the vectors of affective states of the affective and deliberative systems respectively related to these stimuli, h is the effort necessary to correct the desire that comes from the affective system, function of the power of the will, W and elements that weaken it, σ, and M are the courses of action of the affective system.

This model tells us that the deliberative system is subject to two forces: one from the deliberative system itself and another from the affective system. If the first one totally overrides the second one, the behavior followed would be xD, and if only the affective one prevails the behavior would be xA. However, what usually (but not always) happens is that an intermediate point is reached between both extreme positions. And after applying this model to three different problems: intertemporal preference, risk behavior and altruism, they come to the conclusion that the affective system shares the regulation of the behavior with the deliberative system, and that the totally rational behaviors, derived from the deliberative system are not always what we find in reality.

Beyond the simple model of Loewestein and O'Donoghue, it is encouraged to mathematize human behavior in a different way to the neoclassical, this being an alternative modeling direction in Economics; to consider the maximization of both systems (either acting in the form of conflict, as Kahneman argues, or in a unitary way, as Glimcher argues), but not only modeling the deliberative, as has the tradition in economics from the neoclassical to now (and above an unreal deliberative system, arising from introspection, and not from Neurosciences). Perhaps, in some years, we will see many more models with proposals of this type, without too complex mathematics, and probably more refined, both at the micro level and those that support the macro. This is probably the only way for the neuroeconomic approaches to overcome the Friedman Thesis of epistemological validation, that ask for more accurately predictions with less theoretical complexity.

BRAIN, MIND & ECONOMICS

Economics started with Psychology (not scientific) in the middle of the XVIII century, hand in hand with the so-called Classics, such as Adam Smith (especially with his "Theory of Moral Sentiments") and David Hume, among others. To be more rigorous, as the (scientific) psychology still did not exist, those founding fathers of economics acted in some way as psychologists, especially through introspection, recognizing in their writings the remarkable influence of emotions - just as the reason- in the decision making; that is to say the conjunction of both, not the separated reason acting in isolation to make the best decision.

However, a few decades later, the economists who followed Adam Smith and the like - the neoclassical - became skeptical about the possibility that our psychological forces could be measured directly, which led to the adoption by the economic science of the useful tautology between unobserved utilities (originated in the black box of the human psyche) and observed (revealed) preferences.

However, at present, the important advances in Cognitive Neuroscience allow (for the first time in the history of science) to approach a more direct measurement of thoughts and feelings, opening said black box -the human mind-, the basal block of all economic interaction. Undoubtedly, this scientific possibility is disruptive for the Economic Sciences, since it allows facing the generation of economic theory and its subsequent verification / falsification from other methodological paths.

Making a very brief historical synthesis of the traditional micro models of decision making, the Theory of Revealed Preferences (in English it was known as WARP) was developed initially in 1930 by the famous economist Paul Samuelson (and later refined, in what it was called GARP), becoming the core of the so-called "neoclassical revolution". The theory, in its GARP version, proposes that if a consumer, when facing the choice between an apple and an orange, chooses the apple, he is revealing his preference for the apple (without necessarily knowing the neuro mechanisms that led him to that decision -It does not matter to the neoclassical-). Additionally, if the same consumer reveals preferring oranges over pears, this implies that "indirectly" is revealing preferences of apples over pears, that is, which (observed) decisions can be used to predict about the relative desirability of many pairs of other goods, although they have never been directly compared by consumers. Thus, what Samuelson and later authors demonstrated mathematically was that even simple assumptions about this kind of binary choices, revealing stable (albeit weak) preferences, could have powerful implications for economic theory.

After the WARP and the GARP, but within the same idea, came the theoretical refinements of the so-called Expected Utility Theory (von Neumann and Morgenstern) and Subjective Theory of Expected Profit (Savage), always with the sole purpose of predicting decisions (choices), no matter in the least the internal process in the "human black box", your brain. Undoubtedly, and put into context, the contribution of Samuelson and his followers and refiners was truly ingenious; However, today, with

42

neuro advances, it is clearly insufficient to understand the complexity of the economic decision-making process.

But they do not begin with Neuroeconomics the attacks against the "revealed preferences", but they come already from the decade of '50; for example in 1953, the French economist Maurice Allais demonstrated certain "failures" of that theory, which went down in history as the "Paradox of Allais". In 1963 the so-called "Ellsberg Paradox" was added, with dyes similar to that of Allais, in the sense of discovering faults that violated the main axioms of the Theory of Revealed Preferences and all its later versions. Subsequently, during the '70 and '80 came the Nobel prizes Simon, Kahneman and Tversky (mentioned in detail in a previous chapter), with theoretical and empirical contributions that noted that the range of phenomena that was outside the traditional theory of "Revealed Preferences" and "Expected Utility" was much greater than that implied by the paradoxes of Allais and Ellsberg. Of course, at present, neuroeconomists have joined, with instruments of measurement much more sophisticated than previous critics, and each time discovering "more holes in the neoclassical ceiling", more anomalies, which already cause doubts about their true scientific rigor.

1. Has Economics an Unconscious Base?

Sigmund Freud was one of the greatest intellectual figures of the 20th century, an Austrian neurological doctor of Jewish origin. Much of his work remains, to this day, highly controversial, where some point to his as a genius, while others highlight his alleged lack of scientific seriousness.

Freud tries to give an explanation to the way mind operates, proposing a structure divided into three parts: the id, the ego and the superego.

- The ID represents the primal impulses and constitutes the engine of human thought and behavior, motivation and our most primitive gratification desires.
- The SUPEREGO is the part that counteracts the id, representing moral and ethical thoughts.
- The EGO remains between them, and acts mediating between our primitive needs and our ethical and moral beliefs.

The Freudian theory covers several aspects of human psychic functioning, with a high preponderance the Austrian doctor gave to two points: the unconscious and the sensation of pleasure, repressed or not, in the interpretation of human behavior. Let's rescue the following paragraphs of Freud:

"The unconscious is the largest circle that includes within itself the smallest circle of the conscious, all conscious has its preliminary passage in the unconscious, while the unconscious can stop with this step and still claim full value as a psychic activity."

"The unconscious of a human being can react to that of another without going through the conscious."

Consumer Neuroscience today teaches (via neuroimaging techniques not available in Freud's time) that we do not know at all why we choose what we buy. The decision would be taken, to a large extent, below the threshold of consciousness, where our most instinctive biology and our most emotional parts, the Freud's ID, sharpen. The ID would elucidate the interest in the product, the intention to purchase and the loyalty to the brand. These elements correspond to the construction, induced by the promotional campaigns, of desires and brand loyalty. Undoubtedly, the Freud's postulates 100 years ago are not at all far from these modern findings.

The mechanism would work, at the ID, in the following way: certain sensory stimuli (induced by large corporations) activate deep areas of our brain. The reward system (limbic and subconscious), especially the nucleus accumbens and ventral striatum, are put into action and drives to seek food, sex and safety, the three basic pillars of human survival.

Brands, creations of large corporations to induce, seek to activate brain areas that regulate the sense of belonging, making us part of a group, a tribe, a community. All this, together with the natural tendency to imitate and / or empathize with everything that surrounds us (mirror neurons), leads us to consume much less rationally than we believe, pulling down the dogma of free choice, the sovereignty of the consumer.

Therefore, the Marketing of Emotions tries to strongly exploit the Freudian concept of the ID - the most primitive and hidden instincts of the human being, to create value and, ultimately, benefits. Today it seems as a strong resurrection of Freud's ideas, at least under the Consumer Neuroscience field[24].

Dopamine and consumer pleasure center

That is, we live clearly today in a world where, thanks to Neuromarketing, corporations are learning to find product mixes that give maximum sensory enjoyment to the consumer (visual, tactile, auditory enjoyment, etc.), generating a true Freudian Economy, in the sense of enjoyment and pleasure, not repressed this time. The cerebral reward system, around the ventral striatum and the nucleus accumbens (limbic system), where the neurotransmitter king is dopamine, is key in this process.

It turns out that this neurotransmitter influences the sensation of pleasure in the brain, and therefore, shapes the tastes and preferences of consumers. Its secretion increases during pleasant situations and stimulates one to look for that activity, occupation or pleasant goods and services.

Dopamine area is considered as "pleasure center", since it regulates motivation and desire and causes us to repeat behaviors that provide us with benefits or pleasure. It is released with both pleasant and unpleasant stimuli, causing us to demand more of something, or to avoid them if the result is unpleasant. It is very studied also in the case of addictions.

Its objective is clear: to make us want to repeat one or more behaviors, as a way to assure existence. For example, the pleasant sensation we feel when having sex or eating something delicious, makes us want to repeat the action, ensuring the survival of the species through the reproduction and / or consumption of food. That is to say, for Economics, dopamine is of vital importance, being one of the main responsible for modeling the consumer's preference curves, and the whole valuation-pricing system of the economy.

To sum up

Neuroeconomics shows today that the unconscious basis of behavior, highlighted by Freud, connected to the dopamine centers of pleasure or cerebral reward, are not far from the economic reality, and today large corporations are designing real experiences of pleasure for its consumers, generating a truly Freudian paradise of high added value for companies, which at some point will lead governments to assess how much danger they represent in terms of purchase addictions, but that today represent great profits for companies.

2. Some Neoclassical's Anomalies

In 1952, a few years after the publication of the Von Neumann and Morgenstern expected utility theory, a meeting was held in Paris to discuss risk economics. Many of the most renowned economists of the time were present. Among the American guests were futures Nobel laureates Paul Samuelson, Kenneth Arrow and Milton Friedman, as well as the illustrious statistician Jimmie Savage.

One of the organizers of the Paris meeting was Maurice Allais, who a few years later would also receive the Nobel Prize. Allais set out to show that his guests were susceptible to a "certainty effect", and that, therefore, they violated the theory of expected utility and the axioms of rational choice in which that theory rested.

Allais's paradox was later developed by Maurice Allais in his book "Le Comportement de L'homme Rationnel Devant le Risque: Critique des Postulats et Axiomes de L'école Américaine", published in 1953.

In the first bet the least risky option is preferable to a higher expected utility, while in the second bet a higher profit is preferable to a less risky option. That ends up being the paradox, based on the fact that in financial risk or betting choices, although people generally prefer certainty to uncertainty, if the bet is presented differently, they will prefer the uncertainty that was previously rejected.

As Allais had anticipated, the well-educated participants in the meeting did not notice that their preferences violated utility theory until the moment they were reminded that the meeting was about to conclude. Allais demonstrated **that the most**

outstanding decision theorists around the world had preferences that were inconsistent with their own concept of rationality. Apparently, he believed that his audience, persuaded, would abandon the approach that he somewhat disdainfully labeled "American school" and adopt its alternative logic of the election he had developed.

However, Allais was going to suffer great disappointment. The majority of economists, little fans to the theory of the decision, ignored the problem of Allais. As often happens when a theory that has been widely accepted and considered useful is challenged, they saw the problem as an anomaly and continued to use the theory of expected utility as if nothing had happened. On the other hand, the decision theorists (a group we can find statisticians, economists, philosophers and psychologists) took Allais' challenge very seriously. When Amos Tversky and Daniel Kahneman began their work, one of our first goals was to find a satisfactory psychological explanation of Allais' paradox.

Most decision theorists, maintained their belief in human rationality and tried to twist the rules of rational choice to allow this pattern. For years there have been multiple attempts to find a plausible justification for the effect of certainty, but none has been convincing. Amos Tversky was little patient with these efforts; he called on theorists who tried to rationalize the violations of the utility theory "lawyers of confusion", since together with Kahneman they went in a different direction. They maintained the theory of utility as a logic of rational choice, but abandoned the idea that humans are perfectly rational in their choices. They set out to develop a psychological theory that would describe the choices people make regardless of whether they are rational or not. In the perspective theory (prospects), the decision values are not identical to the values of the probabilities.

Fortunately, and thanks to all these strong criticisms over the last 50 years, there is now growing curiosity about Neuroeconomics, Behavioral Economics and other "rebellious" branches towards the neoclassical status quo, although still with uncertain credulity about what can change important aspects of traditional economic theory, the neoclassical. It happens that the tradition in economic science of ignoring neuropsychological regularities in making assumptions, both in the micro and macro models, is so strongly rooted-and in fact has proven to be, to some extent, successful, that to know more about the brain and of its underlying neuropsychology seems to be unnecessary for a few colleagues. And it is likely that economists continue a few years more hesitant to give importance to the new neuro findings, beyond the curiosity that they show today, and that they have also shown with Behavioral Economics; but nevertheless, it is difficult to believe that certain neuroscientific regularities are going to be ignored for a long time, especially those that help explain better certain anomalies that have been discussed for years in our discipline.

Mention some of these anomalies, for example, in order to illustrate possible contributions of Neuroeconomics to solve them. They argue Camerer, Loewestein and Prelec [25], that in many areas of economics there are basic or variable constructs that can be usefully thought as neural processes, and in this way, studied using Neuroimaging, Trasncranean Magnetic Stimulation and other related tools (these tools have already been mentioned in a previous chapter). For example, let's take the field of finance, where millions of daily stochastic observations are made in markets, but despite such statistical access, and after decades of arduous academic research, there is still little agreement on basic issues such as why prices of financial stocks are usually so volatile, based on changing risk perceptions. Perhaps knowing a little more about the neural mechanisms that underlie the assessment of risks by human beings, biases and other human "fragilities" can help explain these theoretical riddles better.

Continuing with the enumeration of anomalies in economic theory, let us now turn to labor markets, where a major question is still why wages are rigid to the downside. It is generally said that companies are afraid of such casualties because they want to keep high the "morale of the workers"; and that paying a high salary also induces effort. But probably, this "workers' moral" is not sensitive only to salary levels, but also depends on the feelings of employees towards their employers, and also can be very sensitive to recent experience, to the opinion of other workers, whether the salary cuts are procedurally fair, among others. And there are no reasons why these aspects cannot be described as neural processes and studied in this way, hand in hand with Neuroeconomics.

Also, within the current theoretical base of economics, there would be an important series of anomalies in terms of intertemporal choices. In the United States, Camerer, Loewestein and Prelec mention, debt with credit cards is quite high at present (about US $ 5,000 average per family) and, as a consequence, a large number of personal bankruptcies are declared annually. There is also the case of low-calorie food, which is cheap and easier to obtain than ever before, but spending on diets and treatments for obesity (no cheap at all) is growing more and more. Surely, understanding how brain mechanisms process reward for what we consume, or how they produce compulsion (shopping, food, etc.), could help explain these facts and shape effective policies on the subject, since analysis based on traditional economic theory (hyper-rationalist) do not fit too much.

But the empirical findings of alleged anomalies crop up everywhere. Let's see additional examples, in this case from the work of the Peruvian economist Ernesto López, which is more based on Behavioral Economics than on Neuroeconomics, but illustrates the current theory-practice disparity in economics with eloquent examples[26]. For example, let's go back to the field of finance and consider investor overconfidence. In theory, rational investors are expected to make periodic contributions and withdrawals from their investment portfolios, which try to keep them balanced in terms of the profitability-risk ratio and carry out some transactions for tax purposes. However, it is difficult that these legitimate needs of the rational investor can justify the high volumes of transactions registered in stock exchanges throughout the world. In a very interesting work, Barber and Odean [27], empirically evaluated the behavior of a sample of 35,000 investors from the United States and came to the conclusion that:

- the volume of transactions was excessive compared to what was recommended and,

- as a consequence of this behavior, agents that carried out the most transactions, in general, obtained worse results than the market average.

Something else: in the same study, investors were classified by sex and it was found that males (who, moreover, are overrepresented in the financial sector worldwide) made 45% more transactions than women and obtained lower net profits by approximately one percentage point, a statistically significant margin.

What explanation can be given to these results? In these cases we speak of overconfidence, which consists of the conviction of an agent, that the accuracy of his knowledge about the value of an action is superior to that of the market and that is reflected in the current price.

In agreement with the empirical findings, psychological studies show an excess of confidence in men with greater intensity than women, especially in what refers to tasks that are perceived as "masculine" - among which finance is counted- and in those situations in which the feedback information is non-existent or ambiguous (again, this is the case of finance). So, even when both men and women show signs of overconfidence, the excess of confidence of the "macho" in an activity that assumes as "his domain" leads him to invest in excess and to obtain worse results than women. That is, again, the neoclassical maximizing cost-benefit calculation seems to fail, and what is worse, we are talking about a large sample of investors, not isolated cases.

Another interesting example is related to household savings. In effect, the theory of the life cycle, widely accepted in the traditional academic world, predicts that people will save during the most productive periods of their lives and will get into debt or consume their savings during the years of lower income. Clearly, this prediction is not supported empirically. On the contrary, it is appreciated that the consumption of people is very closely related to their income and that, in many cases, the consumption of individuals falls drastically when they go to retirement, simply because they do not have enough savings to "soften" their consuming patterns. An analysis conducted for the United States shows that many middle and lower income families simply do not have the capacity to save and, therefore, do not save. And if this happens in the United States, surely similar studies in Latin American countries would lead to results, similar or probably worse.

We can also give as an example the case of those markets characterized by the use of veiled information (hidden): it is verified that there are several markets where companies choose to hide information from consumers. Take as an example banks, which spend large amounts on advertising to express the virtues of their services, but do not sufficiently highlight the various costs that the consumer must assume, such as commissions and expenses of various kinds. In this case, although banks could compete based on these charges (as indicated by conventional economic theory), they decide to hide them, in such a way that most consumers take a long time to understand the cost structure of services associated with their bank accounts. And similarly, in the printer market manufacturers compete intensively for the cost of printing equipment, but they do not compete with respect to the main cost associated with having a printer, namely, ink cartridges only compatible with one type of equipment, that can end up costing ten times the value of the equipment throughout its useful life.

As already mentioned, in these cases, conventional theory would imply that this concealment of information would end up affecting the agent responsible for it, since the veiled information - which is probably not favorable to consumers - would lead to the "rational consumers" discover the information or, at least, establish the conjecture that hidden prices must be high prices and, consequently, be directed towards those suppliers that do not hide information. In balance, all suppliers would reveal the full information relevant to consumers.

However, the results of the analysis show that the existence of "myopic" consumers leads to the emergence and permanence of information hiding behaviors by suppliers, a situation that would configure a market equilibrium in which a part of the information is veiled. These results are consistent with other research that show that consumers give more weight to the sale price of an electrical device than to the cost of the associated electricity consumption during the product's useful life, or that reveal that, in the case of purchases over the Internet, the consumers pay more attention to direct costs than to shipping costs.

Through all these eloquent examples, we have analyzed just a few of all the anomalies that the traditional, hyper-rational theory, cannot explain today, and that "give rise" to the fact that

Neuroeconomics (and also Behavioral Economics) can help to overcome them, with results so far promising. Next, we will analyze more in detail specific findings that different research teams in Neuroeconomics are currently obtaining around the world.

3. The Economic Brain under Risk and Uncertainty

In a classic neuroeconomic papers, *The Neural Basis of Financial Risk Taking*, Kuhnen and Knutson[28] tell us that financial investors systematically deviate from rationality when making their portfolio decisions, and in this way, in their study, they try to identify neural mechanisms responsible for such anomalies. Using fMRI (neuroimaging), the authors examined whether, by anticipating investors' neural activity (i.e. by seeing what goes on inside their brain during decision making), optimal and suboptimal financial decisions can be predicted. They characterized two types of deviations with respect to the optimal investment decision (neoclassical):

- risk search errors;
- risk aversion errors.

As for the concrete results, it was found that activation of the nucleus accumbens (eminently emotional area of the brain, activated when the person has a marked preference for something) preceded both risky choices and risk-seeking errors, while activation of the anterior insula (part of the emotional brain, center of disgust-displeasure) preceded choices without risk and risk aversion errors. These findings suggest that:

- different neural circuits, linked to anticipatory effects, promote different types of financial decisions,
- and that excessive activation of these circuits can lead to investment errors (risk and search aversion).

In this way, they conclude that taking into account anticipatory neural mechanisms can add predictive power to the rational decision model of neoclassical economics, which evidently "remains in shame" in the face of empirical evidence.

More about Risk and Neuroeconomics

People react to risks at two different levels. On the one hand, people try to assess the objective level of risk that different scenarios have. But on the other hand, people also react - in situations with a certain degree of risk and uncertainty - on an emotional level, and such emotional reactions can greatly affect their behavior.

The existence in human beings of separate systems for the cognitive and the affective, which respond differently to the risks, is more noticeable when the two systems collide. People often seem to be "two minds" (one deliberative and one more visceral) when facing situations with risk: for example when we have to invite someone to leave, or speak before a certain number of people, or take an important examination, our deliberative mind uses various tactics to propel us to take risks, which perhaps our visceral (emotional, non-deliberative) mind would prefer to avoid. Perhaps the most dramatic illustration of the separation of visceral reactions and cognitive / rational evaluations is found in the various degrees of phobias that people suffer: what distinguishes a phobia is the impossibility of facing a risk that one recognizes - objectively- be little dangerous (move by elevator, by an escalator, to name some of the most scandalous). Moreover, the fact that we humans spend some money on drugs and / or therapies to overcome our phobias, is a clear sign that our deliberative and visceral systems are not in mutual peace usually.

However, today there is much that is known about the neural processes underlying the emotional / affective responses to risks. Most of the risk-averse behaviors are caused by fear responses / fear of risks, where this fear seems to originate in the region called the amygdala (the center of fear, located in the emotional part of our brain). The amygdala constantly monitors new stimuli that indicate potential threat and responds to inputs from both automatic and controlled processes in our brain. However, the amygdala also receives stimuli from the cerebral cortex (the most rational part of the brain), which can moderate or even eliminate the emotional response.

The decision making under risk and uncertainty, as for example the case of intertemporal elections, adequately illustrate both the collaboration and the competition between the emotional and rational systems that exist within us. The case of the difference in risk taking between people with brain damage in the pre-frontal zone (which produces a disconnection between the emotional and rational systems) and normal people is much cited; the former always tend to make decisions that are much riskier than the latter. And while clearly, having pre-frontal damage to the brain in general decreases the quality of our decision-making, there are particular situations in which people with brain damage such as the above can make higher decisions than normal people, for example before very risky scenarios where normal people are usually paralyzed.

The evidence from Neurosciences also substantiates the distinction between risk (known probability) and Knigthian uncertainty (ambiguity). Different studies with neuroimaging show that different degrees of risk and uncertainty activate different areas of the brain. For example Ming Hsu and others[29] found greater activation of the frontal insula and the amygdala (both eminently emotional zones) when people faced ambiguous choices (uncertainty) compared to risky ones.

Once again, it can be seen that Neurosciences, and specifically, a consideration of emotional and automatic processes - both long forgotten by economists in dominant economic models - could potentially lead an important line of research and theory, argue Camerer, Loewestein and Prelec in his aforementioned paper[30]. And they add that, if the current theory continues failing to incorporate the affective dimensions of risk, it will be unable to shed light on such important phenomena as the ups and downs in the stock markets, the betting markets and the vicissitudes of public responses to threats as diverse as terrorism and global warming, to name just a few important issues.

4. Neuroeconomics and Game's Theory

Game theory is an area of applied mathematics that uses models to study interactions in formalized incentive structures (so-called games) and carry out decision processes. Their researchers study the optimal strategies as well as the predicted and observed behavior of individuals in games. Apparently different types of interaction may, in fact, present similar incentive structures and, therefore, jointly represent the same game.

While economics was one of its first applications (especially for oligopolistic markets), game theory today is used in many fields, from biology to philosophy. It experienced a substantial growth and was formalized for the first time from the works of John von Neumann and Oskar Morgestern, before and during the Cold War, mainly due to its application to military strategy. Since the seventies, game theory has been applied to animal behavior, including the development of species by natural selection. In the wake of games like the Prisoner's Dilemma, in which widespread egoism hurts the players, game theory has been used in political science, ethics and philosophy. Finally, it has also attracted the attention of computer researchers, using artificial intelligence and cybernetics.

But punctually in the field of economics, Neurosciences in general and Neuroeconomics in particular are already well equipped to explore the main assumptions upon which the predictions of game theory rest. These assumptions are:

- players have appropriate beliefs about what others are going to do,

- have no emotions or concerns about what others earn,
- plan forward,
- learn from experience.

In strategic interactions (games), knowing how other people think, and also knowing how other people think you think, is critical in predicting other people's behavior. Nowadays, many neuroscientists think that in the human brain there is an area specialized in "mind reading" (also called Theory of Mind), probably in the pre-frontal zone of our brain, known as area 10 of Brodmann, which generates reasoning about what people who interact with us probably think and then do. In fact, autism is believed to imply a deficit in this area and related circuits. People with autism often have problems imagining what other people think and believe, and therefore are driven to have abnormal behaviors for the common people.

McCabe and others[31] used neuroimaging to measure brain activity when different people played games involving trust, cooperation, rewards and punishments. They found that those players who cooperated showed significant activation in the aforementioned Brodmann area 10 and in the thalamus. On the contrary, those who cooperated little did not show systematic activation in those areas.

Also interesting is the research by Tania Singer and others[32], who reported an important link between reward and behavior in certain games. These researchers, played the participants of their study, repeated games of the type "prisoner's dilemma", where some players, while they were scanned, faced a series of opponents. First, only the scanned participants were informed that some of their opponents would cooperate intentionally while others would cooperate, but unintentionally. Subsequently - also only the scanned ones - they were shown the faces of those against whom they had played. The faces of the intentional cooperators activated the insula, the amygdala and areas of the ventral striatum, among others. And since striatum is a brain area related to rewards, activations in this region meant that simply seeing the face of people who intentionally cooperated with one is retributive.

More about Theory of Games and Neuroeconomics

In an interesting work on the relationship between Neuroeconomics and Theory of Games, the Argentine economist Alfredo Navarro[33] tells us that, apart from the importance that Neurosciences have for Economics -in particular to redefine the rationality hypothesis-, it is also important to keep in mind that there is a mechanism to export economic methodologies to neuroscience and biology, giving a new perspective to the theory of evolution and allowing analyzing the reciprocal behavior of living beings, where Game Theory plays a very important role. That is, according to this vision, there would be a round trip: Neurosciences impacting Economics, which gives rise to Neuroeconomics (the object of analysis of this work), but also, and this is the novelty, Economics impacting on Neurosciences That is, a soft science impacting a hard science. Let's see how this is. In what follows of this section we will make a review of the work of the aforementioned Navarro, which in turn is based on the very interesting work of the neurobiologist Paul Glimcher[34], where this round trip between Economics, Neurosciences and Biology is analyzed.

Paul Glimcher, who comes from the field of medicine, not economics, in a recent work entitled: Decisions, Uncertainty and the Brain. The Science of Neuroeconomics, analyzes the behavior of living beings based on their effect on other living beings and of these on the first, trying to establish a new paradigm for a better interpretation of the behavior of living beings in general and of humans in particular. Glimcher, after reviewing the ideas about the nature of human behavior of Hippocrates, Galen, Harvey, Bacon and Galileo among others, considers Descartes (1596-1650) as the founder of neuroscience. Divide human behavior into two types, the simple and the complex. The first corresponds to the responses to the impulses of the environment, where there is no free will, as when we perceive the heat of a flame near one hand and quickly remove it. This was revolutionary, because no one before had seriously argued that a phenomenon as complex as behavior could be seen as the product of pure physical interactions in physiological systems.

But the complex behaviors have as characteristic that they are at the mercy of the soul, which supposed lodged in the pineal gland, and that can decide freely according to the circumstances. While the first type of behavior is determined, as is the movement of the planets, whose trajectory we can foresee exactly, it does not occur as well as the second, where free will retains all its validity.

The idea that human behavior, at least that which we call simple, was perfectly predictable took more force at the end of the 18th century with the development of the mathematics of Leibnitz, Newton, Lagrange and Laplace, which allow to predict the future position of the planets every time with better precision. Why then not analyze the behavior of living beings with the same purpose of predicting their behavior? Charles Scott Sherrington, an Oxford neurophysiologist, at the beginning of the last century laid the foundations for the physiological study of reflexes, through a neat description of the processes, but still maintaining the Cartesian distinction between simple, deterministic behaviors and complex behaviors, not deterministic. Subsequently Pavlov generalized the analysis of reflexes to the totality of human behavior and therefore also generalized determinism to all human behavior.

Several reactions against the Sherrington paradigm took place, especially that of Marr, who in the seventies proposed a different hypothesis: behaviors should be analyzed in terms of the organism's objective, which is basically to maximize their "inclusive fitness", meaning that rate at which genes are propagated. But to this must be added the fact that living organisms do not have a full knowledge of the world that surrounds them, for which reason they find themselves in a situation of relative uncertainty. The deterministic mathematics, which was the basis of the theories of reflexes, become insufficient, and it is necessary to resort to the mathematics of the uncertain, that is, to the theory of probabilities, since we rarely have a total knowledge of the circumstances around us. Although the theory of probabilities was born in the eighteenth century with Pascal and Bayes, three centuries pass until it is incorporated into human behavior, both in economics and in neurobiology.

In this way Glimcher, through his historical analysis, presents a way to analyze the behavior of organisms from two different perspectives: simple behaviors, in the Cartesian division, can be solved by applying classical economic theory, because either there is nothing random, or the uncertain is due to our lack of knowledge, so we must use the calculation of probabilities. But in other circumstances -complex behaviors-, we must resort to the theory of games, to analyze behaviors that are unpredictable, not because epistemologically we do not reach knowledge to explain the causes of behavior, as Pavlov maintained, but because they are, necessarily, intrinsically random.

This is a very striking statement for two reasons, firstly because it implies accepting that economic theory explains not only human behavior, but the behavior of all beings belonging to the animal kingdom, and not only economic behavior, but all kinds of behavior, and in second term because, to this affirmation, it is not made by an economist, but by a neurobiologist. According to Pavlov and Laplace, the uncertainty comes from the lack of knowledge of who decides, while what Glimcher says is that the uncertainty comes from outside, from the outside world to who decides, and that the latter must necessarily make a random decision if you do not want your opponent to predict your behavior and gain an advantage from it.

In this way, following the reasoning of the neurobiologist Glimcher, the analysis of the behavior of living organisms can be understood much more fully if we do so from the perspective of game theory, which we remember begins to be applied to the analysis of economic problems with the appearance of the developments of von Neumann and Morgenstern, in 1944, where non-cooperative zero-sum games are analyzed, but more especially after the Nash developments, which analyzes the determination of equilibrium in more generalized situations, such as games cooperatives and non-zero sum.

The analysis of the behavior of organisms that have brains allows Glimcher to argue that there are two types of uncertainty: one that we can call epistemological, which is originated in the lack of information and knowledge of the agent, and that could allow a mechanistic interpretation of the behavior, and another that derives from the need to follow a random behavior.

Suppose a lion is in front of a lamb. You can jump to the right or to the left, trying to guess the behavior of the lamb. Suppose that it can also jump to the right or to the left. If it jumps in the same direction as the lion, it is lost, but if it does it in a different direction, it can be saved. If he always jumped in the same direction, the lion would know in advance what his behavior would be, and he would always be lost. But if he tossed a coin into the air to make his choice, he would be saved, for example, 50% of the time, all on condition that the lion does not know in advance what he is going to do. Therefore, random behavior is essential to pursue what has been defined above as "inclusive fitness".

In this way, the mentioned Glimcher reaches its conclusion[35], in the sense that:

We should begin to employ probabilistically based approaches to understand how the brain takes information from the outside world and uses that information in concert with stored representations of the structure of the world to achieve defined computational goals. It has been my central thesis that this goal can be best achieved through the synthesis of economics, biology and neuroscience. The central challenge facing neural scientist is to link behavior and brain...

Economics was designed to be just that, a mathematical corpus which attempts to describe how any goal should be achieved in an uncertain world like the one we inhabit. Behavioral ecologist recognizes this; their field is focused on the study of how animals approximate economically defined goals with regard to the maximization of inclusive fitness.

Experimental economics recognize this; their field is focused on the study of how economic behavior approximate economically defined goals with regard to the maximization of utility. Neurobiologist are also beginning to recognize this, and today it seems natural to assume that some form of Neuroeconomics will play a critical role in explaining how the brain of humans and other animals actually solve the maximization problems this two other disciplines have identified.

In short, Alfredo Navarro, in his great review on the work of Glimcher, illustrates us about something that should fill us with pride to who we come from a soft science such as economics: we are in a position to export analytical tools to tougher sciences such as neurobiology, since it has been discovered that, for example, Game Theory, is a very useful resource to understand the behavior of a large part of living beings, and not only of companies in their economic interactions (such as the theory of the oligopoly).

Understanding the economic mind of others

In a truly leading study, Sanfey, Rilling, Cohen and others[36], tried to determine in two different games (Prisoner's Dilemma and Ultimatum), if people who interact socially, receiving feedbacks from other human beings, and intuiting how these feedbacks could be used to infer how our brain works, could predict what others think. Recall that in game theory, one of the most important tasks for participants is to act strategically from what others do or plan to do, and this implies a key role of the so-called Theory of Mind, i.e. those circuits' brain cells that are activated when trying to predict the behavior of our interlocutors.

The so-called "Theory of Mind" studies our social brain. One of the distinctive attributes of human social cognition is our propensity to build models of other minds, that is, to make inferences about the mental states of others. This human capacity has become known in Neurosciences as a theory of the mind and many neuroimaging studies have attempted to elucidate the neural substrates of this natural human ability. Previous studies to the here detailed have already shown the main activable cerebral areas (some more rational, others more emotional) in this type of action.

The brains of the participants in this experiment (led by the aforementioned Sanfey) were scanned using fMRi (functional magnetic resonance) while playing two different games: Ultimatum Game (UG) and Prisoner's Dilemma (PDG), both in front of other humans and in front of computer screens. Comparing both games, a striking degree of coincidence was observed between the brain areas that were activated, including both areas already accepted as specific to the Theory of Mind (mentioned above), as well as several other brain areas that had not been previously reported, and that may be related to the immersion of participants in real social interactions. And while the interactions of humans with computers also achieved activation in some of the same areas activated by games between only humans, in the latter case these activations were more notorious and defined.

In both games, the participants witnessed a decision on the part of their partners, in the UG they observed an offer of money that another made them, either fair or unfair, and on which they had to react and in the PDG they observe an election what another did, whether cooperative or selfish, and about which they also had to respond. That is, before deciding the answer to take, in both cases, they witnessed something that revealed the partner's intentions. What brain areas would be activated in both cases? That was the central core of the study.

If in the previous study the activated brain areas were analyzed when responding to a fair or unfair offer, in this new study[37] the previous moment was analyzed, that is to say, the activable brain areas when a proposal was recently known, just or unjust, and it is deliberating what to do, and at the same time, inferring what the other person is like and his true intentions.

Going to the concrete results of the study, for both games (UG and PDG), activation was detected in two

70

of the four classic areas of the Theory of Mind: anterior paracingular cortex and posterior superior temporal sulcus (STS later). Both areas were activated in interactions with both humans and computers, but showed stronger responses to human partners in both games, that is, respondent participants rejected unfair offers from humans to a greater extent than from computers in the UG and cooperated more often with humans than with computers in the PDG.

Following with the results of the study -where we remember there is social immersion of the participants-, brain areas were also found that were activated that had not been noticed in previous studies -without social interaction-. These were:

- precuneus
- upper temporal sulcus (STS) medium
- an area that includes hypothalamus, middle brain and thalamus
- left hippocampus

Both the activation of the posterior cingulate and the hypothalamus can be related to emotional issues when receiving responses from humans, who obviously have less presence when doing studies without human interaction. The activation of the average STS, normally attributable to the biographical memory, may be related to the fact that the participants are learning new information about other people -the ones who make the offers-. Finally, the activation of the hippocampus could be related to the activity of decoding behaviors and intentions of others: are they just or unjust? Are they cooperative or non-cooperative?

In summary, and taking into account that the paper leaves perhaps more questions than answers, the brain areas that can be activated with respect to the theory of the mind (many of them more emotional than rational, without a doubt), would be at least:

- the anterior paracingular cortex
- upper posterior temporal sulcus (posterior STS)
- the posterior cingulate / precuneus
- the average STS
- an area that includes hypothalamus, middle brain and thalamus
- the left hippocampus

Other Studies

In other landmark study in Neuroeconomics, Sanfey, Rilling, Cohen and others[38], applied fMRi (functional magnetic resonance) about nineteen players of the Ultimatum Game, to investigate the neuro fundamentals of the cognitive and emotional processes put into play when making economic decisions. The aforementioned Ultimatum Game (in this case a single shot -one shot game-) consists of two people trying to share a certain sum of money: one player proposes a division and the other can accept it or not.

Brain images were taken only of the players responding to the proposals (not those who formulated them), where such formulated proposals were sometimes fair and sometimes unfair. The offers considered fair (50/50 distribution of money, or half for each) were all accepted, while unfair offers (all those involving a distribution below 50/50 for the respondent) were more rejected as that increased their degree of injustice (60/40 is not the same as 80/20). And through the neuroimages, it was observed that these unfair offers activated brain areas related to both the emotional (anterior insula) and the cognitive (dorsal-lateral pre-frontal cortex). And in another data that is interesting, it was also observed that the degrees of rejection of unfair offers were greater when the bidder was a human being than when it was simply the computer (who were also used in this experiment as formulators of proposals), illustrating that human beings have a superior emotional reaction to unfair offers from other humans than to the same formulated via some impersonal mechanism (computers in this case).

Another interesting finding of this work was given that, in the face of unfair offers that were later rejected, greater activation of the insula than pre-frontal cortex was observed, while the accepted offers showed the opposite, greater activation of the prefrontal cortex than insula. This situation would be reaffirming what is already known in Neurosciences: the rational / cognitive tendency of the pre-frontal cortex and the eminently emotional nature of the insula. But beware... it is not a competition in our brain between the rational and the emotional separately, but it is a performance of both together, related and complementing.

Also, in another interesting finding, it was observed that the activation of the pre-frontal cortex remained constant before less or more unfair offers, perhaps representing how stable the mental representation of a monetary maximization is, while the activation of the insula scale depending on the degree of injustice of the offer.

Finally, Sanfey and other researchers also observed, in the case of unfair offers, an activation of the anterior cingulate, a cerebral area bordering the pre-frontal cortex, normally activated in situations of conflict between the emotional and the cognitive, such as this one experiment.

In this way, we can conclude that the observed activation in the anterior insula (eminently emotional area of the brain) before unfair treatment or offerings, indicates a very important role of emotions in human decision-making processes, despite the attempt of the standard economic theory for suggesting that any sum of money offered to a person - without any cost or consideration - should be accepted, since net income is maximized.

To sum up

In general, all these neuroeconomic papers, combined with game's theory, suggests that the human being does not always maximize in his economic decisions, since sometimes, although the economic calculation advises one clear path, the emotional influences, making the decision apparently irrational, taking other way. But such decisions are not irrational, are just human.

5. Oxytocin, Trust and Market Economies

No one can argue, surely, that trust between people is essential to strengthen human societies. Trust is necessary to make friends, form partners, families and organizations and of course play an essential role in economic exchanges and politics. In the absence of trust between people and companies, market transactions are cut, and in the absence of trust in the institutions and leaders of a country, political legitimacy is lost. Recent empirical evidence in humans has identified the role of neuroactive hormones, especially oxytocin, as a facilitator of pro-social behavior based on trust.

Recent neuroeconomic experiments with humans have shown that the reception of a signal of confidence from a stranger is associated with an endogenous release of oxytocin by the brain and also that high levels of oxytocin have been strongly associated with reciprocal behaviors of said signals of trust.

In this work, Paul Zak and Ahlam Fakhar[39], test whether the endocrinological bases of trust between humans (in small groups, that is, at the micro level) can be scaled at the country level (macro level), especially taking into account the statistics on confidence at the national level show substantial disparities (in Norway for example, 65% of respondents answered that they could trust their fellow citizens, while in Peru only 6% responded in that way).

Oxytocin (a type of neuroactive hormone we said), whom Zak calls the "molecule of morality", is synthesized in the hypothalamus (belonging to the limbic system - eminently emotional zone of the brain) and then released into the circulatory system. In humans, certain areas of the brain associated with memory (the diagonal band of Broca and the basal nucleus of Meynert) and areas associated with emotions (hypothalamus and amygdala) present an important accumulation of oxytocin receptors, although there are receptors of oxytocin distributed throughout the brain. This distribution of oxytocin receptors in limbic areas suggests that the decision to trust others has an important emotional component, and therefore a high component of speed and low introspection when deciding.

And, as both studies with animals and humans, indicate that estrogen is highly related to oxytocin levels, the authors of this work used estrogen as a proxy for oxytocin. The hypothesis to be demonstrated in this study was that people who live in societies settled in environments with high levels of oxytocin and / or estrogen are more likely to affirm that their fellow citizens are reliable, that is, to have more confidence in their peers.

Analyzing in detail the work, thirty-one variables were taken (between biological, social and environmental) associated with interpersonal trust for a sample of forty-one countries, where the authors found that two groups of variables are related to trust interpersonal at the country level: the consumption by its inhabitants of plants based on estrogens (phytoestrogens) and the existence of environmental conditions that include the presence of molecules of the estrogen type. In this way, these results provide preliminary evidence that levels of confidence at the country level may be related to the intake of neuroactive hormones by its inhabitants, via food or via the environment, mainly.

They also comment Zak and Fakhar that there are more than 300 plants in the world that have been identified as phytoestrogenic. For example, phytoestrogens are found in foods such as soybeans and derivatives, rye and derivatives, rice, beans, beef and tea / mate, among others.

In summary, this paper shows that endocrinological effects can be a new explanation-independent of the usual institutional causes-for the problem of confidence differentials observed between countries, indicators directly associated with higher or lower levels of investment and economic development of each country. That is to say, this work tries to show that specific environmental / food conditions in some countries, which impact the oxytocin levels of its inhabitants, can lead to higher levels of confidence. Specifically, nations that have high per capita incomes, clean environments and consume more food with phytoestrogens have a good chance of showing high levels of generalized trust among their inhabitants, which facilitates economic transactions in general and investment levels in particular.

This information, Zak and Fakhar conclude, should be useful for politicians, if they are interested in raising the levels of trust among their governed, and therefore the quality of their market systems, especially in developing countries. Also the conclusions of this work give a certain rationality towards the maintenance of clean environments and towards the consumption of healthy foods.

6. Libertarian Paternalism as a Behavioral Public Policies Guide

Why are there so many people who smoke a lot or are addicted to different types of drugs? Why do so many people eat junk food in excess? And more generally ... why do so many people voluntarily decide to do things that they know hurt them in the long term?

Richard Thaler, the last Nobel Prize in Economics, and solid member of the "Economics of Behavior" School, argues that the problem originates in the limited rationality of human beings. In their mental processes, argues the academic, people separate the immediate effects of an action from the aggregate and long-term effects of it, valuing them in different ways (usually more value to the present than to the future), and behaving systematically in a contrary to their own benefit.

In this way, Thaler justifies the state intervention, "libertarian paternalism" he calls, to remedy the incorrectness of people with an exacerbated "limited rationality", giving them a nudge in the right direction. It is, without a doubt, a form of interventionism that liberal libertarians will blaspheme forever.

In any case, the intervention suggested by Thaler is much less "interventionist" than those we are accustomed to seeing in the real policies of the day-to-day governments. Thaler argues that, given the imperfect and limited rationality of many people, small changes in the rules of initiation could encourage people to behave in the "socially desirable" way, reducing long-term interventionism. For example, the basic rule, for him, should be the donation of organs after death; if someone did not want to donate, they could opt for it. The junk food must be in the most hidden places of the supermarkets, so that the effort of buying it is greater. If someone does not manifest their willingness to have a pension fund, it must be considered that they do want one.

Origins of the idea

The term "libertarian paternalism" was coined by the aforementioned behavioral economist Richard Thaler and the jurist Cass Sunstein, in a 2003 article in the American Economic Review. The authors developed their ideas in a more extensive article at the University of Chicago Law Review that same year.

In the aforementioned article, they propose that, both from the private sector and from the government, it is about influencing the behavior of people to make their life longer, healthier and better. They continue that, in proven findings of the social sciences, it has been shown that, in many cases, individuals make very bad decisions, decisions that they would not have made if they had paid attention and had had all the information, unlimited cognitive abilities and absolute self-control.

And while it is paternalistic / interventionist, they justify that it is liberal / libertarian in the sense that its goal is to ensure that people are freed from many of their biases of limited rationality, to disassociate from disadvantageous agreements, if they prefer. According to them, libertarian paternalists want to facilitate people to follow their own path; they do not want to put obstacles in the way of those who wish to exercise their freedom.

Critics

Thaler's critics say that his "libertarian paternalism" is just a modern justification for state interventionism, which starts from considering people are irrational because they do not make the decisions that a certain group of people find desirable.

And critics add: if people are really irrational, as Thaler says, what makes sure that those who design the rules are not? What assures us that their "pushes", far from helping us to be better, enslave us to their tastes and appreciations, depriving us of our tastes and our appreciations?

Justification of Thaler

Libertarian paternalism is a relatively weak and soft type of paternalism that does not involve interference, because the options are not blocked or eliminated, nor are they taxed in a significant way. If someone wants to smoke, eat a lot of candy, subscribe to unfavorable medical insurance or not save for retirement, libertarian paternalists do not force him to act differently, they only induce him with incentives.

Thaler and Sunstein argue that government and private companies often become "architects of choice", because our perceptions often depend on how we organize the different options that are presented to us. The world is full of these "architects" - parents, religious leaders, professors, doctors, etc. - who influence our choices and have the responsibility to give them shape through "nudges", which do not limit us but can compensate for human error, if we use them correctly.

In summary

State interventionism based on "socially desirable patterns" is not new, although the term "libertarian paternalism" can be. The idea of inducing through "incentives" (fiscal, or otherwise) certain economic behaviors, which do not arise spontaneously, by limited rationality or for whatever reason, is the guide of modern economic policy for a century at least, although now it is better grounded in neuropsychological terms.

Therefore, new labels for old uses and customs of economic policy, although this time with a bias towards a more limited, more intelligent interventionism, since it is based on a deeper knowledge of human rationality, backed in Neurosciences, and not in mere philosophical speculations about the human psyche, quite deficient in many cases.

EMOTIONS & ECONOMICS

Neoclassical model basically supposes an economic agent with a single decision center[40], the deliberative (the more rational part of the brain), to explain the economic behavior of the consumer that maximizes its utility, the businessman who efficiently organizes his company, the offender who faces to the risk of being arrested if he commits a crime, and of the one who makes the decision to marry or have children, to mention a few examples.

But Neuroeconomics, as we have commented until the satiety in this book, confirms that there are two decision systems: **the affective and the deliberative.** The first corresponds to the internal parts of the brain, that is, the most primitive in the evolutionary stage, and the second is located in the cerebral cortex and appears in more recent stages of the evolutionary process. The affective system is related to emotions that have effects on the motivations of human behavior, with a value component always present, either biological (fear, hunger, sexual desire) or social (sympathy, hatred, distrust), and usually operates in meta-conscious form.

The deliberative system, on the contrary, acts by evaluating what the affective system perceives, with which it is bound by biunivocal nervous connections, and over which it exerts a certain power by having its will power to correct the behavior that would be followed if it existed only the affective system, as it happens with the most primitive animals. The stimuli can affect the affective part only, or also the deliberative part, and depending on the evaluation of both systems, the behavior to be followed will be defined. With these assumptions, which are the basic contributions of Neuroeconomics, Loewenstein and O'Donoghue go a step further, to build a mathematical model that allows them to formalize this relationship. They assume that the human being faces a function to be minimized, which is the cost of his behavior. A part of the cost is the difference between what the deliberative system wants and what it ultimately obtains and another part of the cost is the effort that the deliberative system must make to turn the impulse of acting in a certain way.

$$[U (x_D, c (s), a (s)) - U (x_A, c (s), a (s))] + h (W, \sigma) [M (a_A, a (s) - M (x, a (s))]$$

where U is a utility function, x the chosen course of action, of a set X, the supra-indexes D and A indicate the optimal behaviors for the deliberative and affective systems respectively, s is a vector of stimuli, y_a (s) and c (s) are the vectors of affective states of the affective and deliberative systems respectively related to these stimuli, h is the effort necessary to correct the desire that comes from the affective system, function of the power of the will, W and elements that weaken it, σ, and M are the courses of action of the affective system.

This model tells us that the deliberative system is subject to two forces: one from the deliberative system itself and another from the affective system. If the first one totally overrides the second one, the behavior followed would be x_D, and if only the affective one prevails the behavior would be x_A. However, what usually (but not always) happens is that an intermediate point is reached between both extreme positions. Subsequently, the authors apply this model to three different problems: intertemporal preference, risk behavior and altruism. In all three cases, they come to the conclusion that the affective system shares the regulation of behavior with the deliberative system, and that totally rational behaviors, derived from the deliberative system, are not always what we find in reality.

In *The Vulcanization of the Human Brain* [41] , Cohen, renowned American neuroscientist, considers human behavior in terms of its evolution from more primitive forms, in which the cerebral cortex did not yet exist. He considers that the brain is a confederation of mechanisms, which sometimes act together, but at other times they compete with each other (other neuroeconomist, as Paul Glimcher, don't think the same). Cohen describes an experiment in which the behavior of different people is analyzed in the face of the dilemma of avoiding the death of five people by sacrificing a sixth. When the decision must be made at a distance from the facts, we use to accept the suggestion of the cold rationality, avoiding the death of five at the expense of the death of the sixth. But when immersed in the problem, close to the facts, the limbic part of the brain seems to have priority (eminently emotional zone), and we are reluctant to sacrifice that sixth person.

The author, who analyzes the emergence of the human cerebral cortex rationality as something evolutionary, attributes it to the fact that our ancestors did not have the possibility to act at a great distance, due to natural danger the were exposed. The cortex, which would have been the result of a process of vulcanization of the brain, has generated a technological system that has exceeded our emotional capacity. It is a very complicated task to produce a nuclear device, but it is very simple to press a button to throw it. This could imply that the evolution of the human being has led to a crossroads difficult to solve because the cerebral cortex has developed, capable of enormous progress that perhaps would not have occurred in the limbic brain, and that would mean, in that case, and that evolution has taken a bite of the apple of Eden.

And to finish with the synthesis of neuroeconomic papers made by Alfredo Navarro, let's analyze now the work entitled Damage to the prefrontal cortex increases utilitarian moral judgments, where its authors (Michael Koenings, Liane Young, Ralph Adolphs, Daniel Tranel, Fiery Cushman, Marc Hauser and Antonio Damasio[42]) analyze whether emotions play a causal role in ethical judgments, and how different areas of the brain contribute to that end. They analyze the behavior of six patients who present lesions in the ventromedial prefrontal cortex, (a region of the brain necessary for the control of emotions, and particularly of social emotions), which have an extremely utilitarian behavior when deciding on moral dilemmas. This type of work illustrates the way in which damage to the brain can be an alternative way of studying its functioning.

Most of the studies that are being done in Neuroeconomics to date, aim to identify the components of rationality and emotionality that are behind of each economic decision, defying rational expectations and similar models currently in force.

In short, Neuroeconomics is unleashing a theoretical discussion "that brings them", and that surely will reach high decibels in the coming decades, no doubt.

1. The Emotional Memory and Economic Decision Making

The somatic marker hypothesis, developed by Antonio Damasio in 1998, is a theory that has been very relevant when understanding the role of emotion in decision making. It is argued that before the consequences of a decision there is a certain emotional reaction that is subjective, that is, it can be "experienced", and at the same time somatic, that is, it is translated into muscular, neuroendocrine or neurophysiological reactions. This emotional response in turn can be associated with consequences, whether negative or positive or sets of stimuli that define a situation, that are repeated with certain constancy over time and that provoke such response. This mechanism of association is what produces what Damasio calls "somatic marker" and that is defined as: "a bodily change that reflects an emotional state, either positive or negative, that can influence the decisions made at a certain time". In this way, it is stated that the emotional reaction goes from being a mere consequence, for example of some negative decision, to influencing the decision making itself, making possible the anticipation of the consequences and guiding the final resolution process.

In this sense, it is affirmed that somatic markers can provide unconscious (or, in general, metaconscious) signals that "facilitate and contribute to decision-making", even without the subjects being able to explain the reason for their strategy (for example, when we buy products that clearly would not suit us from the point of view of the cost-quality ratio, or when businessmen manifest, at certain times, an aversion to risk that seems irrational). One way to study these "somatic marker" effects is shown by the IGT (Iowa Gambling Task), where the word gambling gives us the idea of tests based on bets / games of chance, in which different people must choose between four heaps of cards, and depending on their choice they receive rewards or symbolic monetary punishments, so that in the long run two heaps will lead the participants to lose while the other two win.

These tests -in their majority- have been carried out through the study of the changes in the electrodermal activity (skin conductance levels and response) produced by the decision-making situation. For example, the works of Bechara - another prominent neuroeconomist nowadays - have shown that normal subjects show greater cutaneous conductance responses when the probable consequences of their choices - gains or losses - are greater. However, the greatest wealth of this research lies in the finding of anticipatory electrodermal responses, that is, they appeared just before the subjects made the choice. The researchers observed that the subjects who chose the heaps of cards with the highest profits showed a greater conductance response before choosing the disadvantageous decks (with lower gains), which has been interpreted as an anticipatory corporal signal that guides the subjects avoid said deck.

These conclusions, together with those of other studies carried out in recent years, have placed the pre-frontal cortex, especially the ventromedial orbitofrontal portion, in the "key region" for decision-making, since it is in this zone where the consequences of long-term actions are evaluated, thanks to the integration of somatic states with information specific to the situation and with stored memories of similar situations.

Lavin concludes that these findings have supported the idea that there are anticipatory somatic responses (supported and reinforced by experience) that guide future behavior and the choices made in similar situations, positioning somatic markers as a relevant variable to consider when evaluating decision making and the relationship between it and emotion. This is also reinforced by the differences between the electrodermal responses of people who achieve optimal performance in the development of tasks and those who achieve poor results and also those who have neurological damage in the brain areas involved in these responses.

Another interesting study related to the subject is one of Natalie Denburg[43], which analyzes the influence of somatic markers through the different stages in a person's life, especially discriminating between adults under and over 70 years.

Her paper analyzes the correlation that exists between decision making and people's psychophysiological responses, where through the use of the (already mentioned) Iowa Gambling task (IGT), it was observed how they intervene in adults greater changes in skin conductance (SCRs) compared to the anticipation generated prior to the behavioral response. It was found that older adults who obtained negative results in the test (that is, low performance in the bets) presented greater discrimination of the changes in the conductance of the skin during the execution of the test (IGT), reason why it was concluded in which decision-making ability is subject to the interpretation of somatic markers, and that the reduction in decision-making capacity in older adults (over 70 years) would be due to an abnormal functioning of the somatic response against the anticipation of future events.

It was observed that decision making is assisted by emotional processes, signs or somatic markers that originate not only in the body but also in cortical and subcortical areas, including the ventromedial prefrontal cortex (VMPC), the amygdala and the insular cortex, somatosensory cortex, the basal ganglia and the peripheral nervous system. Needless to say, many of the areas mentioned belong to the so-called limbic system, that is, our emotional, meta-conscious and non-rational brain.

In another very interesting aspect of the study, it was also found that, during the game, older adults produced greater amplitude of psychophysiological response to the decks that offered greater reward and that younger adults generated greater amplitude of psychophysiological response to the decks that offered economic disadvantage, which suggests that the pattern of anticipatory discrimination during successful IGT performance differs for older adults and younger adults.

This translates into the fact that decision-making effectiveness in older adults is due to the anticipation of the positive, as opposed to the young adults who sustain their decision making in anticipation of negative responses.

Somatic markers can be positive or negative, although in both cases, they are vital for the action. If positive somatic markers prevail in decision-making in older adults, this suggests that positive somatic states promote approaching behaviors and that younger adults use negative somatic markers to evade options that do not suit them.

But why this difference between young adults and older adults? Some researchers argue that older adults would have greater attenuation of their emotions by experiencing less negative affect than their younger counterparts, presumably for existential reasons[44].

Summarizing then the concept of somatic marker, and beyond the two particular cases analyzed in this section, we can conclude that:

- the somatic marker is "learned" in past experiences;
- the somatic marker is noted in situations in which certain current events are associated with past emotions;
- when the marker is reactivated, either consciously or metaconsciously, the partial or complete replication of an emotional state associated with the current situation to be resolved is promoted;
- the somatic marker, like "memory trace", is recorded in high-order cortical circuits, of which the ventral-medial prefrontal cortex is the most notable example.

In summary, we believe that the hypothesis of the somatic marker, by the hand of Damasio, Bechara and several other researchers in recent years, promises to be of great importance to unite the theory of hyper-rational decisions (on which economics is based on the neoclassical school) with the reality observed in the business world, where very important decisions are made every day, which involve a lot of money, but where the emotional, the intuitive and the metaconscious sometimes end up giving priority to the final decision, improving the success of it, and not the other way around.

✓ The Intestine Intelligence

"Thinking with your guts" is not a simple expression, since there are several neurological and psychological processes that were believed to take place exclusively in our heads, but which actually originate in our intestines.

For a long time it was believed that the intellect was the ability, housed in the brain, to use reason to learn and know. Hence the importance post industrial revolution of reason, the scientific method, and the hypothetical-deductive, and other scientific herbs. But now, XXI century, we know that our digestive system is also key to our reasoning and our behavior, but of course ... not being at all our intestines calculating and logical-deductive. It is the so-called second brain, or brain of the bowels, and has an important influence on our economic decisions.

The second brain is composed of about 500 million neurons prowling the digestive system, which represents five times the number of neurons found in the spinal cord. Its main function is to regulate the motility of the intestine, but also to influence our perceptions, our sensations, our reasoning, and above all ... our decisions.

This organ is full of serotonin (the chemical that regulates our mood), much more than the brain of the head. Approximately 80% of this neurotransmitter is produced and found in the intestine.

Therefore, many studies today analyze the correlation between the health of the intestine, which depends on good bacteria, with mental health, because bacteria also interact with the central nervous system, that is, the brain of the head.

The first brain

It is the best known (of the head), and is composed of neurons and glial cells, in the form of networks, communicated thanks to neurotransmitters. The more complex neurological functions such as learning, memory, economic decision making, among others, depend on the ability of these cells to form fortified neural networks.

The second brain

It is formed mainly by the neurons of the walls of the intestinal tract, can operate autonomously and usually communicates with the first brain through the parasympathetic nervous system (the vagus nerve connects the intestine with the brain of the head) and the sympathetic system (the prevertebral ganglia).

In something that is important, the communication between both brains is completely bidirectional: the first brain influences the functioning of the intestine, but also the neurons that make up the second brain influence the functioning of the first. These bowel-brain signals are a powerful influence on emotions and behavior, especially in response to disturbing or threatening stimuli and events.

Prepare our second brain to make better economic decisions

At this stage of the 21st century, science is clear that economic decisions, to a great extent, are predefined goods before they reach consciousness, influenced notably by the limbic system (the center of the emotions of the first brain), and the intestines (the second brain).

People before the supermarket gondola, investors in the stock market, politicians with macroeconomics, and absolutely all human beings who make some kind of decision are notably influenced by unconscious mechanisms located both in the limbic system and in the brain of the intestines, being key to what we finally end up deciding with our scarce resources, be it money or time.

In this way, it is necessary to be at peace with our intestines, not sending signs of stress, for which experts recommend performing meditation or yoga to reduce the levels of cortisol, the stress hormone that directly affects our digestive system.

Another factor to consider about intestinal intelligence is that the intestine is the habitat of thousands of bacteria: a way of life that forms its own ecosystem in this long organ. Taking that into account, it is important to respect our biological clock (circadian cycle), because if we alter them we will also be altering the life of the bacteria and their work in the intestine, and, therefore, also our humors and our decisions.

In that sense, specialists recommend eating many probiotic foods - microorganisms good for the body, such as soups, yogurt, certain breads and fermented foods. Probiotics regenerate the intestinal flora, or what is the same, balance the ecosystem of bacteria in your digestive system.

In this way, good habits such as meditating, sleeping as necessary, eating well, generate healthier people, but also more intelligent to manage their economy, their money and their cost-benefit calculations.

To sum up, the somatic marker hypothesis and the ideas of the second brain are just a few examples of a new trend to understand economic decisions from its roots, the emotion, not the reason.

2. Anxiety, Emotions and Time's Economic Perception

Economists define intertemporal decisions as those with consequences over multiple periods of time, including a wide range of decisions, of varying degrees of complexity and frequency, such as investments in real and financial assets, savings for retirement, purchases with credit cards, purchases of merchandise for the home in advance, etc.

✓ The Traditional Theory

To study and model intertemporal decisions, traditional economics has generally used the theory of discounted utility, based on the idea that economic agents prefer a similar reward more if it is obtained in the present than in the future; and similarly, future costs would be less painful than the costs to be faced today.

To formulate these theories, models have been generally used based on the assumption that the total utility of a series of rewards and / or costs over time can be decomposed into a weighted sum (or integral) of utility flows in each period of time.

A particular case is the function of exponential discount, which has as its main characteristic that the discount rate is independent of the passage of time, and consequently, the evaluation of a course of action towards the future (project) is independent of the moment in which the project is analyzed. This property of decisions is called dynamic consistency.

However, the problem with this exponential model is that it cannot explain several empirical regularities, that is, it would be incompatible with reality in certain cases. In fact, several field studies show that discount functions decline at a faster rate in the short than in the long term, that is, people are more impatient when they make short-term exchanges (today vs tomorrow), than when they make exchanges in the long term (day 100 vs day 101).

To illustrate, the empirical evidence suggests that if a person is given $ 100 to choose now or $ 110 tomorrow, he may prefer $ 100 now, while the same person may choose $ 100 in two years or $ 110 in two years and one day, he could prefer $ 110 in two years and one day. It seems then that discount rates tend to be higher in the short term than in the long term.

✓   Alternative approaches

At present, some economists familiar with the neuro have studied alternatives to exponential discount functions. The generalized hyperbolic function has the property of declining at a higher rate in the short term than in the long term, adjusting the cases of inconsistent decisions. Ainslie (1992) [45] and Loewenstein and Prelec (1992)[46] have used this type of function in their studies.

Another highly studied discount function is also the quasi-hyperbolic discount function, which also captures the property that the short-term discount rate is high and the long-term discount rate is low. The quasi-hyperbolic equation is generally referred to as the function biased to the present and was first proposed to model the planning of the transfer of wealth between generations, and then applied to an individual scale by David Laibson (1997) [47] in the model of the golden eggs to study intrapersonal financial decisions.

These models would better capture the dynamic inconsistency of the preferences, that is, the idea that the passage of time changes the preferences of the agents and, consequently, projects that can be positive evaluated with some initial time perspective can be turn negative if they are evaluated from other time perspective.

Also models of dynamic inconsistency have been used to study problems of self-control: credit card expenses, drug addictions, etc.

✓ Neuro Fundamentals

As noted above, the quasi-hyperbolic function of discounting time provides a good fit to experimental behavioral data, however few studies have focused their analysis on identifying the causes of this tension between short-term and long-term preferences. Then the following questions arise:

- What is the mechanism behind these intertemporal decisions?
- Do they arise from a single preference mechanism or from multiple systems that interact?

In an attempt to answer these questions, Samuel McLure, David Laibson, George Loewenstein and Jonathan Cohen [48] , using functional magnetic resonance imaging (fMRI), examined the neural correlate of time discount while subjects made choices between monetary reward options that varied across time.

The experiment consisted of giving the participants a choice between a sum in the short term and another in the long term, the first being less than the second. Both options were separated by a minimum time lag of two weeks, and in some pairs of options, the earliest option was immediately available.

The hypothesis was that the behavior pattern of the two parameters (β and δ) arises from the joint influence of different neural processes. The β related to the limbic system and the δ related to the lateral prefrontal cortex and other structures associated with higher cognitive functions (the more rational ones).

What results did the researchers obtain? Basically, there would be two systems involved in such intertemporal decisions:

- parts of the limbic system (emotional zone of the brain) associated with the dopamine system of the central brain, including the paralimbic cortex, which would be triggered by decisions involving immediately available rewards;
- regions of the lateral prefrontal cortex and the posterior parietal cortex (eminently rational areas of the brain), uniformly involved in intertemporal decisions independently of the delay in time.

This neuroeconomic finding is consistent with the evidence that consumers act impatient today but prefer to act patients in the future, also supporting the hypothesis that different neuronal systems are activated by intertemporal decisions: the impatience of the short term, which is driven by the limbic system (emotional, not deliberative), and that responds preferably to immediate rewards and to a lesser extent to future rewards; and long-term patience, dominated by the lateral prefrontal cortex and associated structures (the most deliberative parts of our brain), which can rationally evaluate exchanges between abstract rewards, including rewards over longer periods.

Finally, we believe that future research should better assess what kind of discount functions are ideal for predicting real-world economic decisions, and generally improve methods for measuring intertemporal decisions, where Neuroeconomics will undoubtedly play an important role.

✓ Devices to Calm the Economic Anxiety

Beyond our natural intertemporal preference for the present more than for the future, there is a growing consensus among intellectuals (of all types) of an acceleration of nowadays way of living, with so much people in an anxiety induced by the marketing of the big corporations, that exploit our system 1 (quick decisions) via a product differentiation that theoretically will make us happier, but on the contrary, it is making us work many hours, to earn more money, and thus buy those panaceas from happiness, happiness that never ends up consummating, because we never reach the necessary money to cover all those "induced desires". And so, we find today a truly epidemic of people anxious, with panic attacks and anxiety disorders, depression, etc.

And although in recent years, companies have emerged that reorient their strategic policies, implementing programs of "B Companies", among others, with the aim of producing in a more responsible way and re investing in society and its workers, the ultimate solution passes by an internal transformation of the human being, in order to see what can truly achieve welfare in a sustainable way (which is very far from the current induced consumerism), and in that yoga, mindfulness, and related devices, can be very useful.

Mind Relaxing, Innovation & Creativity

The term "mindfulness" suggests preparing the mind to pay full attention to what is happening, what we are doing, and the space in which we are moving. That may seem trivial, except for the fact that we very often deviate from the issue at hand, we get lost in the multitasking and the frenetic daily rhythm. In this way, our mind takes flight, we lose contact with our body and, very soon, we are absorbed in obsessive thoughts about the immediate, what has just happened, seeing everything threatening. And that makes us anxious, accelerated.

Mindfulness helps as an antithetical force of the problem, training the basic human capacity to be fully present, aware of where we are and what we are doing, and not being too reactive for what happens around us in the immediate.

The "full consciousness" is a natural quality that we all have. It is available to us at all times, but only if we take the time to appreciate it. When practicing mindfulness, say the specialists, the art of creating a space for ourselves is practiced: space to think, space to breathe, and in general a space between us and our reactions.

The idea is to take a few minutes a day to quiet our mind, trying to pay attention to the present moment, without judging it. Our mind will wander. As we practice paying attention to what happens in our body and mind in the present moment, we will see that many thoughts arise. Our mind will try to deviate towards something that happened yesterday, to take us out of balance, as a kind of "wandering mind".

But the "wandering mind" is not something to fear, it is part of human nature and provides the key moment for the practice of mindfulness: the piece that researchers believe leads to healthier and more agile brains: the moment we recognize that the mind has wandered. Because if we can notice that our mind has wandered, then we can consciously return it to the present moment. And the more we do this gymnastics, the experts say, the more likely we will be able to do it over and over again.

It is about returning our attention again and again to the present moment, without judgments about our thoughts. Our minds are wired to be carried away by thought. That is why mindfulness is the practice of returning, again and again, to the breath. The sensation of breathing is an anchor to the present moment. And every time we return to our breathing, we reinforce our ability to do it again.

Innovation and creativity are ignited with mindfulness, a fact that is no less important for entrepreneurism in the economy. As we deal with the increasing complexity and uncertainty of our world, mindfulness can lead us to effective, resilient, low-cost responses to problems that are difficult to solve in traditional consciousness planes.

The key to a better personal and economic well-being goes through a spiritual peace that you have to work to find. The path of emotional marketing that of large corporations, and their induced consumerism, is probably not the most appropriate. Instead, mindfulness, among other techniques, paves the way we need to reflect deeply on where we want to go, find ourselves ... and see clearly what is really important to get there ... leaving the rest as secondary, separating the straw of wheat.

To finish, the context is not going to change, hyper connectivity and Neuromarketing of large corporations will be increasingly powerful, so we are the ones who have to change, to adapt and be happier in this complex world, redirecting our economical temporal preferences toward an equilibrium of body, mind and soul.

3. Basal Ganglia and Aversion to Change: Why Keynesian Find Fundamentals in Neuroscience.

Is it good to overprotect companies and institutions to endure? According to the renowned economist Nicholas Taleb, overprotection instead of helping to do something stronger, on the contrary, makes it weaker. Like the overprotected children, those organizations that are deprived of elements of stress, in the long run they become weaker. And although the argument seems indisputable, it has an important neuropsychology contradiction, which we develop below.

For Taleb, antifragility goes beyond resilience or robustness. The resilient resists the blows and remains the same, whereas the antifragile becomes better with the blows. Antifragility becomes strong with randomness, uncertainty, the volatile, the unknown, the incomprehensible and the errors. "What does not kill me, makes me stronger", would be a good phrase to exemplify Taleb's concept of antifragility.

According to Taleb, organizations (private, public, NGOs) should tend towards less interventionist patterns, where the natural takes its course, and especially, randomness. For him, the environment of organizations is much more complex than what our memory or historical account can tell. In addition, the educational system and the scientific apparatus would be designed, in their vision, to organize all events in a linear manner, simplifying too much.

In this way, although in nature anti-fragility is the norm, the scientific story rejects antifragility, often preventing interferences with things that it does not understand, confusing the unknown with the non-existent. Impossible to disagree with Taleb in that regard, but of course ... scientists are human beings, flesh and blood like everyone else, whose brains seek assurances and regularities, which limit environmental uncertainty. That is the flaw we notice in Taleb's argument, and on that point is where we want to go deeper.

In the human brain, there is a region that participates notably in the formation of routines and habits, since acquiring a routine requires considerable effort, the brain stores in its memory the "template" of the habit, to reactivate it at the slightest sign. These patterns are developed and established in the so-called basal ganglia brain region, whose functions are essential in the acquisition of habits, addictions and learning processes.

Habits help us in our daily lives, because they allow us not to have to decide each of our actions continuously, and thus reduce the consumption of energy in the brain. Constant routines are thus delimited before we get down to work, which saves us time. In this way, the natural thing is that the brain tends more towards the routine than towards the disorder and change, which implies that it has to be trained to achieve what Taleb proposes (in fact, many international companies today pay onerous training for this type of programs for its executives).

In this way, habits arise because our brain is always trying to save mental effort: these fragments of automatic routines are stored in the basal ganglia, so that when we execute an automatic routine, "work" the basal ganglia, and the rest of the brain "rest", simplifying a little. In other words, the permanent change Taleb poses is unnatural to our brain, unless we train it. The tendency is towards order, rigidity and permanence, and not vice versa; and in the end our institutions are the result of our brain format.

Following with his notion of fragile-robust-antifragile, Taleb establishes some comparisons. Curiosity is antifragile, and books have the ability to multiply it. The banking system is fragile, but Silicon Valley, with its permanent innovation, is anti-fragile. Food companies are fragile, restaurants are antifragile. The bureaucrat is fragile, the entrepreneur is antifragile. A person who depends on a salary to live is very dependent on their organization and very fragile for their level of dependence. An artist is antifragile because of his independence from an employer.

In this way, Taleb criticizes all types of state interventionism to save sectors and companies in decline, since that undermines the mechanism that generates anti-fragility, necessary for the system to innovate and be increasingly productive. But of course ... the argument collides with a reality: our brains have a natural tendency towards aversion to change, that is, towards what the renowned economist calls fragile. Such a contradiction!

In addition to our natural tendency towards routine and inflexibility, dominated by the basal ganglia in the brain, there is a whole question of errors in our decision-making process, depending on the uncertainty of a complex world (which Taleb describes well), and also of our limited ability to analyze all available information (limited rationality of Simon), which enhance the aversion to change of the average human being.

In general, Neurosciences seem to indicate that our brain is not designed for outstanding performances in relatively complex decisions, including the brains of people who have studied at the university level. It is known that the human brain has been developing for millions of years, but for most of our history as a civilization it served for people who only covered basic needs: find food, reproduce and defend the territory, not much more than that. It was not until the last 200-300 years that the world became exponentially complex, which has implied the need for refined neural connections for decisions that are increasingly risky and / or uncertain, but also a good part of the primitive remnant. Our brain's emotional state has remained almost unchanged.

Paul Glimcher, one of the most reputable neuroeconomists today, argues that the valuations (preferences) we assign to objects and actions would be learned by trial and error, where the dopaminergic neurons of our mid-brain (reward system) would play a fundamental role, through the concept of the reward prediction error (the difference between the expected reward of a given course of action and the one actually achieved), which would be limited by learning. That is, the brain predicts, and is wrong, generating errors, and while they are decreasing with experience, in a world as changing as today, learning is increasingly continuous, and errors too.

That is, there are at least two natural tendencies in the opposite direction to Taleb's antifragility: the action of the basal ganglia (our natural tendency to routines and rigidities) and the computational problems of our brain to predict / decide without mistakes (bounded rationality), in an increasingly complex and changing world.

To conclude, and although we agree with what Taleb raises about the importance of freedom and the search for risks to make robust and anti-fragile people, institutions and economics as a whole, it is a difficult proposal to apply for mental models humans have today, and that, in the end, determines our behavior.

The question is why people, institutions and society tend to fall into rigidities, routines and interventionism, which hinders the dynamism that Taleb mentions as necessary for success, innovation and productivity. And the answer goes through our brain: the human being seeks assurances, low risks, conventions and other rigidities that reassure him and give him certain equilibria that flexibility and permanent randomness do not give him.

In the background, our brain functioning is the result of thousands of years of adaptation of human life to the environment where he lives, with a prominent role to the basal ganglia to build routines and other rigidities that reassure us in the face of so much systemic uncertainty. Adapted brains for permanent change are the minority (and you have to train them), the norm is the routine.

To sum up, Taleb's antifragile concept has a deep contradiction: it raises certain libertarian conducts as desirable for our modern societies, when our brains, human nature and institutions have adapted to the world doing just the opposite, creating a Keynesian world of rigidities, which moderate the increased environmental uncertainty. The natural tendency of the human being is to seek order, and not disorder; our brains are wired that way, no other way round. That is perhaps the main reason why governments intervene so much into economics, to temper the changes, because our brains seek gradualism, not shock, although that disgusts so much to fanatics of the free economy.

4. Money, Emotions & Neurofinances

Neurofinances usually perform brain scans while people doing activities where they earn or lose money. The results obtained suggest that earning money activates reward areas similar to those that are activated through the consumption of food and drugs, which would imply that the money confers direct utility, instead of being valued only by what can be bought with it.

The standard economic model assumes that the utility of money is indirect, since it is only a means to facilitate the exchange of goods and services, which are those that end up providing utility directly. Thus, the traditional neoclassical economics conceives the pleasure of food or cocaine, for example, and the "pleasure" of obtaining money, as two totally different phenomena.

However, neurological evidence [49] suggests that the same dopaminergic reward circuits of the brain are activated for a wide variety of rewards, including attractive faces, funny cartoons, cultural objects, sports cars, drugs, and money). So, according to neuro evidence, it would seem that money, like the other goods and services mentioned, provides a direct reward.

Therefore, the idea that many types of reward (whether by buying goods and services or simply by having money in your pocket, even if it is not spent) are processed in a similar way in the brain, has important implications for economics, that he assumes that the marginal utility of money depends on what it can buy; in this way today it is hypothesized that money would become what psychologists call a "primary reinforcer", which means that people would value money without carefully calculating what they plan to buy with it. And while, we acknowledge, Neuroeconomics is not advanced enough today to categorically affirm this hypothesis, there is a very high possibility that brain valuation for money is loosely linked to the utility of consumption.

But then, if earning money directly provides pleasure, the experience of saying goodbye to him will probably be painful. This would be one of the reasons why many consumers tend to accept purchases in medium and long term financing, to disguise the payments, and in this way reduce our pain by getting rid of liquidity.

The Emerging Field of Neurofinances

In an effort to seek better foundations on financial decision making, and using the current boom of Cognitive Neuroscience, a modern area of study has emerged, the Neurofinances, where the use of magnetic resonance images, with people during real risky investment situations (or bets), is essential.

For example, today Neurosciences show that the circumstances that accompany a decision to bet / invest money are seldom independent of the investment / bet itself. The oscillations of stock markets, or the constant variability of betting games (horses, casino, etc.), have a significant impact on the amount of risk that people are willing to take, increasing it.

In this way, observations by brain scanner indicate that the increase in risk taking in bets / investments would be correlated with the emotional reaction caused by being an individual immersed in a situation of volatile bets, which makes the reflective-deliberative activity decrease.

Additionally, and also from Neurosciences, today it is quite clear that the financial choices we make are influenced by our previous experiences, forming what are called somatic markers. The hypothesis of somatic marker, owed to the Portuguese neurologist Antonio Damasio, proposed a way to explain how emotions affect when making complex decisions (including financial investments).

According to Damasio, our previous experiences make us store in the brain a series of sensations (muscular and hormonal responses) pleasant or unpleasant related to certain stimuli. This relationship between stimulus and emotional state is what is called a somatic marker. Faced with the task of making an investment decision in a context of risk or uncertainty, a stimulus similar to that of previous experiences would trigger in our body the release of a certain somatic marker.

Thus, the options we opt for are those associated with pleasant somatic markers, and we avoid those that the somatic marker associates with adverse results. This process greatly accelerates decision making, being a kind of shortcut for the brain, in order to avoid investing excessive deliberative resources in making decisions that require a rapid response. Needless to say, that in matters of financial decisions in stock markets (or in casinos and places of betting), immediacy is permanent, and these cerebral shortcuts, then, become very useful.

✓ Loss Aversion

An interesting concept, from the hand of Nobel Laureates Kahneman and Tversky, explains the concept of aversion to loss, i.e. the idea that the losses of an amount x make us proportionally more damage than the happiness produced by the profits of that same amount x. That is, it would be more than proportional the pain for the loss of $ 500, than the happiness for winning $ 500.

Financial losses are processed in parts of the brain that are responsible for the pain network. One of these areas is the amygdala (the center of fear). For example, patients with this area damaged are proven not to be afraid of losses and often take high financial risks, which normal people do not.

✓ Irrational Investors?

There are many aspects of life in which we make mistakes in making decisions, even sometimes very rude. It is proven that, even in areas where we have some experience, we often stumble from time to time, almost inexplicably. According to many economists, in the world of financial betting (stock markets, casinos, etc.), such repetitive errors are usually abundant, which fuels the debate on how rational are investors when designing their portfolios.

In general, Neurosciences seem to indicate that our brain is not suited for outstanding performances in complex financial decisions, including the brains of people who have studied finance at the university level. It is known that human brain has been developing for millions of years, but for most of our history as a civilization, was adapted for people who only covered basic needs: find food, reproduce and defend the territory, not much more than that. It was not until the last 200-300 years that the world became more complex in an exponential way, arising, among others, financial decisions in securities markets, which has implied the need for refined neural connections for increasingly risky decisions, but also, a good part of the primitive remnant of our brain has remained almost unaltered.

Some think that if some investors are too optimistic and others are too pessimistic at the same time, the market should be able to find its middle ground, compensating, and tending towards rationality on average. However, the empirical evidence in Finance seems to show that individual investment errors tend to move in the same direction and also occur more or less at the same time, that is, it is not that irrational investors would be losing money against arbitrageurs more rational, but that the errors, in certain situations, would be generalized (remember the financial panics, with herd behavior, which so often does not rationally justify similar market collapses, or similar bubbles).

That is, it would not be that investors are irrational because they do not know how to calculate future costs and benefits (the most deliberative part of the brain) in controlled situations (a university exam, for example), but that such calculations would cloud, in practice, in environments that are too volatile and risky, generating excessive herd and panic behavior among investors, caused, to a large extent, by brains overly dominated by their more primitive roots (the most emotional parts).

✓ The Brain Predicts

It has also been discovered that the brain works with predictions, contrary to the previously accepted theory, that it reacts to sensations it picks up from the outside world. In this way, human reactions would be just the adaptation of the body to the predictions that the brain makes, based on the state of our body the last time it was in a similar situation (somatic markers).

In this way, the brain tries to find out what a certain sensation means and what is causing it (for example, a strong downturn in the stock market), to then define what to do with it, and thus build thoughts, feelings, perceptions and decisions, that arrive just when it is necessary, and not a second later, but of course, with errors of prediction and biases, which make such decisions may be unsuccessful several times, until the learning improves the perception, and the subsequent action.

And in this process, it is the limbic tissue (the emotional one) that would dominate, and then direct those predictions to the cortex (the most rational part of the brain). For example, when a person is told to imagine a red apple in their mind, limbic areas of the brain send predictions to visual neurons and cause them to interconnect and fire in different patterns so that the person can actually "see" that apple red The reader can change in this example to red apples for papers that are quoted in the stock market, the conclusions are the same.

In this way, the investors of the market are the architects of their own experiences (somatic markers), which start from the emotional brain, since the limbic regions of the brain send but do not receive predictions. Therefore, our brain would be built so that things work in the reverse of popular knowledge:

it is not "seeing to believe", but the other way around, "believing to see". That is, the perceived risk of certain private securities (corporate risk), or of certain countries (sovereign risk), would be mental constructions that arise mainly from the emotional part of the brains of market investors (that is, their limbic parts), explaining in large part the reason for so many financial panics, bubbles, overreactions and mistakes in general, so common in modern stock markets.

✓ Unconscious Financial Rationality

Psychologists, from many decades ago, have been finding more and more mental operations that operate outside of consciousness. And while the historical tradition says that the unconscious is primarily the repository of repressed thoughts about violence and sex, modern neuro findings speak of a remarkable unconscious rationality, which would help us make quick decisions of a certain quality, being key to Economics and Finance.

Unconscious operations underlie many of our inferences and judgments, as well as a large number of decisions and problem solving, including monetary / financial. In fact, the unconscious mind can often do a better job of these things than the conscious mind, which is no small thing for an economic science traditionally thought of as hyper rational, or at least of rational expectations.

The story goes that the controversial Sigmund Freud held two initial views on the unconscious, cognitive vision and dynamic vision. The first would be what we now call unconscious rationality, increasingly accepted scientifically, while the second would be the famous vision of psychoanalytic repression, with various criticisms from current science. In this way, unconscious rationality would also be based on Freud, although it was not what made him famous.

This unconscious rationality, little studied even in Economics, is clearly connected with the theory of the emotional system 1 of Kahneman, Nobel in Economics, a fast system that allows us to decide on a day-to-day basis with enough survival success, although not necessarily great optimality. But it also connects with the hypothesis of the somatic marker of Damasio, which theorizes about the emotional traces in our long-term memory, putting together a triad of concepts with broad implications for our ideas on Economy and Finance, especially for everything that is known on problem solving, decision making and economic behavior in general. It turns out that our unconscious rationality would be key in the interaction between creativity and problem solving, and between intuition and analytical thinking, among other fundamental concepts to understand how current, highly volatile, globalized and complex financial markets work.

The moods of the market

Going to Economics, the humor or volatility of the markets is generally analyzed as a risk rate, in terms of decisions with random results, and is modeled mainly with statistical graphs that calculate means and deviations in time series. Volatility, in the background, is a measure of the frequency and intensity of changes in the price of an asset, and the greater the volatility, the greater the risk of losing, but also a greater opportunity for high profits.

However, the expected profits of the companies do not vary so much daily by their business models per se, but the volatility of the stock markets comes more than anything explained by the perceptions of their operators with respect to issues outside the contributing firms. , such as the public policies of the countries, electoral issues, exchange rates, etc. This volume of information, so changing and probabilistic, is ultimately the one that forces investors to re-analyze scenarios every day, quickly, appealing largely to intuition and everything called system 1 of Kahneman, much more than the definitive and long-term analyzes of Kahneman's system 2, the slow and sapient.

Neural patterns to decide quickly

Empirical evidence today shows that the unconscious mind, in volatile contexts, such as the financial one, is usually superior to the conscious mind, in order to survive, by learning some types of highly complex patterns, which the conscious mind cannot process quickly. In fact, the rational unconscious mind can learn really complicated patterns, without bothering to inform the conscious mind of its achievement. The important thing is to solve problems, get out of the way, remember the brain does not necessarily seek the truth, but to survive.

Let us not forget the unconscious consumes 90% of the total energy of the brain, which notably helps the total energy economy of the central nervous system, since it achieves fast and correct ways of many crucial day-to-day decisions, which if they are too rationalized consciously, they would spend much more energy. Basically it is a matter of neuronal productivity, it is cheaper in energetic terms to solve problems with system 1 (fast and rational unconscious) than with system 2 (slow and rational conscious), and in the world of high finance it shows even more.

To sum up

As volatility operators, investors / gamblers are notoriously influenced by brain determinants that operate below the threshold of consciousness (nucleus accumbens, amygdala, anterior insula), which in the end are the ones that pre-mold the final decision, under a luck of unconscious rationality quite useful for day to day, with an excess of use in very changing contexts, such as the stock market.

In this way, the 21st century would be giving revenge to the rational unconscious, since although since the twentieth century it had been proposing that most psychological processes are not conscious, the "unconscious" that reached the popular imagination and curricula university students was mainly the irrational and repressed Freudian unconscious, that of wild and sexual impulses, barely controlled by conscious and reflexive reason.

For Economics and Finances, the greatest advantage of understanding the rational unconscious will be to take full advantage of our inner emotional vision, to improve our decisions on investment projects, stock market portfolios and public policy design, among others.

✓   Inflation as a Brain Construction

Rivers of ink have been written on the subject of inflation in the last 70 years. They range from the most liberal versions to the most Keynesian, all in some way theorizing about how such a "sustained and

119

generalized rise in prices" would be generated. Undoubtedly, the vast majority of theories give a preponderant place to the exaggerated increase in the amounts of monetary money, which is why the (serious) central banks take care not to fall into these excesses.

However, this consensus, this matrix, is not necessarily the underlying explanation of inflation, but a resulting one. Result of what the reader will ask? Inflation is the result of our mental models, formed from very young, that support this monetarist explanation of inflation, which is a result from our brain being educated to understand the temporary excess of currency as something negative, to which we must react accepting that the prices go up.

But, and here comes the awkward question... what would happen if the brains of the economic agents had been educated differently, in such a way to perceive the temporary excess of currency as positive, useful for business to prosper, so that the credit was more abundant and people could buy more goods and services, and the economy would flow better. The reaction to the increase in the currency would probably change. Recall that Cognitive Neuroscience shows that reality does not exist in itself, but is a recreation and interpretation of the mind, individual, or related groups, which makes the brain "believes to see" and not "see to believe".

In this way, entrepreneurs, trained in the monetarist matrix, are ahead, that is, at the slightest news of real rise in the amount of money by a central bank, prices rise, and responding to what dictates its somatic marker, emotional and instantaneous, derived from his mental model (matrix) monetarist. However, if their mental matrix were different, they would probably not do that, but perhaps they would increase the supply of product and hours of work, a phenomenon that has been observed in small communities with local currencies, where their increase has not been neutral, that is to say, has had effects on the product, instead of on prices.

And here I stop because I know that raise these issues is to open an unwanted tap, since it is full of unscrupulous politicians who have always wanted to abuse the management of money supply for issues that have nothing to do with the welfare of people, but for its perpetuation in power, and corruption. But, from the point of view of the Applied Neuroscience to Economics, I cannot stop raising the issue of what mental models mean in our behavior and our actions, and that it is proven that the human brain "believes to see" and does not "see to believe", leaving room for another socially shared mental model, the reaction to the handling of the creation of currency could be completely different from what is done today.

In short, the issue raised here is one of those Pandora Box that nobody wants to open, especially by the groups that maintain the control mechanisms of the current matrix (US dollar), which know perfectly that you only need "believe" so that a currency has demand, without needing too much support in gold, hard currency, or anything similar.

Concluding

The results of the studies commented throughout this chapter are only some of the many that have been published in recent years, on the actual functioning of human decision-making processes, both in economics and in finance; and it is our main intention to refer them to show how these tools work, that we have available since a relatively short time, and which promises advances that we cannot predict yet, but we believe could become important and disruptive to what has traditionally been done in Economics.

As Colin Camerer says [50], in some aspects the contributions will be incremental, in other radicals, with Neuroeconomics advancing at an accelerated pace, continuing the knocking down of the postulates of Jevons 200 years ago:

"I hesitate to say that men will ever have the means of measuring directly the feelings of the human heart."

We must get out of the Friedmanian comfort of the "irrelevance of assumptions" and go to challenge everything we have been saying so far, where some economic and financial postulates will remain almost unchanged, and others will change, but what is certain is that many economists, in the next few years, are going to study how the brain and our rationality really works, until convince themselves that the maximizing models that we have been using up to now are extremely limited, in light of the new empirical evidence provided by Cognitive Neurosciences.

5. Reason and Emotions: Dual or Single System?

Paul Glimcher, a main neuroeconomist at present, from his laboratory at New York University, has made numerous experiments and collected a huge empirical evidence about studies done in other parts of the world, which has allowed him to condense all this material into an interesting theoretical model, published at initial version in 2009[51], then actualized, where it is hypothesized about the true functioning of the human brain when making decisions.

The model, which in turn is a brief summary of much what it is known in Neuroeconomics, is called "two-stage", because on the one hand, the assessment aspect of decision alternatives is analyzed (the utility in economics), and on the other, the concrete decision to be taken is analyzed, that is, the reason for the selection of a single one (among several alternatives) and its subsequent execution.

For example, in the ASSESSMENT STAGE, the model describes the neuropsychological circuits through which human beings value the alternatives A, B, C, D and E that we have for a given course of action (for example where to go on vacation next summer), that is, something similar to the utility (the economic concept) that we give to each alternative in the neoclassical model; whereas in the DECISION STAGE, the model describes in what way (brain circuits that are activated) we end up choosing the "supposed" best alternative, say A, to go on vacation.

THE ASSESSMENT STAGE has been studied in more detail and depth in recent times, not so much the DECISION STAGE, which in humans is a bit delayed (but not in other mammals, such as monkeys), mainly due to the fact that (in humans) the temporal dynamics of the selection and execution of a given course of action today is difficult to follow via neuroimaging (fMRI).

In any case, the aforementioned "two systems or stages", VALORATION and DECISION, would not be watertight behavior, since there is some empirical evidence that some characteristics of our valuation process (our preference function) are intrinsically attributable to mechanical processes linked to the decision stage.

In another interesting feature, today a high number of studies shows that certain areas of the ventral striatum and the frontal cortex "learn" and "represent" valuations (preferences) even when "learning" is passive, that is, even when the person is not faced with an action or specific object on which he has to decide.

Reward and Subjective Value

The values (preferences) assigned to objects and actions would be "learned" by means of "trial and error", where the dopaminergic neurons of our mid-brain would play a fundamental role, through the concept of the reward prediction error (the difference between the expected reward of a given course of action and the one actually achieved), an error that would be narrowing down more and more thanks to the aforementioned "learning".

THE DECISION SYSTEM involves large portions of our parietal cortex, among others; that in turn receive direct and indirect projections from the areas of the ASSESSMENT SYSTEM, and, once the decision has been made, the process is projected directly towards the MOVEMENT CONTROL AREAS, for the concrete execution of the decision.

In a fact that is quite limiting for those theorists on welfare issues, in Neurosciences today we know a lot about the neuronal circuits involved in the aspects of evaluation of alternatives, only one little about making concrete decisions, but almost nothing about the neural circuits that act in what is called a person's "sense of well-being"; since as we all know, not necessarily the fact of consuming (even though there is a sharp process of weighing rewards and punishments, or costs and benefits) leads us safely to a feeling of well-being.

In something that is very important, in Neuroeconomics the concept of "subjective value" (VS) is proposed, but in cardinal form, instead of the traditional concept of "utility" of the traditional theory, which is ordinal;

The VS, in this way, being cardinal, is measured in terms of the rate of "firing of neurons" -neuronal firing rates- that occurs in certain areas of the brain before the perception of each object or alternative action to choose (for example the options A, B, C, D and E to go on vacation), where the researchers analyze said "neuronal ignition" from the scan of our brain, via neuroimaging.

The choice of the final alternative when making a decision (the alternative A to go on vacation), would be given after comparing the relative VS between the different options, after a "fouling" of the process by "noise".

The "reward prediction error" -RPE- of a chosen alternative would be given by the difference between the expected VS and the VS obtained when making the decision (for example, alternative A to go on vacation); and through the delimitation of said RPE is how our brain would improve its rating system, in this way, it is getting less and less wrong.

The empirical evidence (and working hypotheses) available today suggest that two brain areas seem to contain all the neurons required to extract VS for any object and action: the ventral striatum (member of the limbic-emotional brain) and the middle prefrontal cortex, and in particular the ventral striatum for actions and the middle prefrontal cortex for objects.

But one thing is the extraction of SV (that is, granting value to options A, B, C, D and E before making the decision) and another one its storage (once the decision to choose A has been made), the purposes of being used in subsequent decisions.

In this way, the SV calculated in the areas mentioned in the previous item would be stored in a much wider area than the ventral striatum and the middle prefrontal cortex, which we had seen almost exclusively involved when SV is granted for the first time to an option. Which would lead to the conclusion that when an SV (already stored) is represented in our brain (for example, when deciding where to go on vacation next year, not this year), it would reflect activity in areas such as the lower frontal sulcus, the insula, the amygdala, the posterior cingulate, the superior temporal sulcus, the caudate nucleus, the putamen, and the dorsolateral prefrontal cortex, and obviously the ventral striatum and the middle prefrontal cortex; that is, a much wider area than the participant in the initial assessment of the option.

However, and in what is a current limitation of Neuroeconomics, the details (i.e. the how, not only the where) of this assessment process -assignment of VS to objects and actions- are just beginning to be understood, since they are difficult to reach via neuroimaging.

Decision Stage

Going to the DECISION STAGE, and as we said at the beginning, it is much less studied than the STAGE OF EVALUATION, always speaking of human beings, not of other mammals, like monkeys, where the empirical evidence is much greater.

In the DECISION STAGE, the neurons of the lateral intraparietal area (LIP) would seem to play a fundamental role, since they would be responsible for representing the relative VS of each decision alternative (the A, B, C, D and E of our example of holidays). Remember that the VS of each alternative comes from the ASSESSMENT STAGE, and arose basically from the neuronal activity of two specific areas: the ventral striatum and the middle prefrontal cortex; but in the DECISION STAGE, the absolute VS of each alternative decision would be transformed into relative VS, and this would occur first in the posterior parietal cortex and then be represented in the LIP area.

As in the ASSESSMENT STAGE, in the DECISION STAGE there is also internal brain "noise", which affects the quality of decision-making. At a certain moment, the set of available options (A, B, C, D and E, with their respective absolute and relative VS) converge to a single alternative, the one chosen (alternative A), which would occur when collicular neurons they exceed their "trigger threshold".

In what is a very important current limitation, it should be mentioned that the majority of these studies on the DECISION STAGE revolve around monkeys, and in particular decisions made through "generation of movements through the eye", which is not the only possible alternative to generate movements. However, always according to Glimcher, there is some empirical evidence that this type of brain structures would also operate for decisions on more abstract objects than those that a monkey can usually choose (and which are more usual in humans); and less evidence that it would also operate for structures that generate movements other than the eye, in both monkeys and humans; clarifying that the "lesser evidence" available is temporarily, especially with the advances that are coming in neuroimaging.

Single or Dual System?

To finish with this impressive model, and as we said at the beginning, Glimcher polemizes with Kahneman and affirms that the output of the ASSESSMENT STAGE is not only input of the DECISION STAGE, but also the reverse path would be observed, since there would be numerous decision circuits interconnected with important areas of assessment, such as the aforementioned frontal cortex and basal ganglia; that is, the process would not be linear or additive, but rather more complex, but unitary.

In fact, Glimcher, at the end of the exhibition of his model, attacked fellow neuroeconomists and behaviorists, such as Nobel Prize winner Kahneman, Laibson or Mc Lure, who proposed the existence of

128

two relatively independent systems that would regulate decision-making, one associated with the emotional (the limbic area) and the other more rational (some of frontal and parietal cortex).

To be more specific, Glimcher criticizes the "multiple ego" rationality models, which generally describe the area comprised by the basal ganglia and the prefrontal mid cortex as an emotional module, which interacts (additively) with a second system organized around the posterior parietal cortex and the dorsolateral prefrontal cortex, which would form a rational module.

The mentioned Glimcher indicates that, for example, it would be relatively proven (in monkeys) that neural activity in the posterior parietal cortex (eminently rational) would predict preferences (supposedly generated in emotional areas), under all the conditions that have been studied (immediate reward, future reward, large and small rewards and rewards of high and low probability).

And later Glimcher mentions a lot more empirical evidence, that together, they would be showing a structure globally involved in valuation activities (STAGE OF ASSESSMENT) and not a structure managed exclusively by emotionality.

Of course, concludes Glimcher, the emotions truly influence our decision-making, especially in the ASSESSMENT STAGE, but in no way would there be "multiple selves", that is, the emotional on the one hand determining valuations (utilities) of objects and actions, and the rational on the other side, deciding which is the best option and giving the order to execute.

And here it is convenient to cite the criticism of Kahneman[52], the Nobel Prize in Behavioral Economics, who does not believe that the evidence cited by Glimcher is conclusive to invalidate the argument that decision-making emerges from a conflict between emotions and reason; the opposite of the "unitary" system proposed by Glimcher.

In fact, always according to Kahneman, there would be important behavioral evidence (more grounded in psychology than Neurosciences) about the existence of "multiple selves" in our psyche, and the importance of conflict; however, he concludes that more empirical evidence is needed from Neurosciences to define the winner of this debate; that is to say, it does not attack in definitive form against the Glimcher model, which is logical, since the evidence in Neuroscience is superior to the psychological one.

Finally, Glimcher acknowledges that there are still important aspects to better specify in his model, basically due to lack of empirical evidence, especially in the DECISION STAGE, since we remember that Neuroeconomics is just touching the decade of life and can still be improved a lot plus the instruments available to open our "black box".

In summary, the neuropsychological system that sustains our decision making would seem to be a "little bit" more complex than the simplified version of neoclassical economics, based on ordinal utility curves, faced with the restriction of the income of each consumer, to be able to determine what quantities are consumed of each good and service, deriving from this model the respective demand curves of each of them.

Undoubtedly this neoclassical model, which is simple and unreal, has been enormously useful for doing science, as we will see in the next chapter, where we analyze whether Neuroeconomics could imply a paradigm shift or not. However, through this model of Glimcher, we have been able to appreciate that today we can measure (via neuroimaging) the true utility that each person obtains from each good or service, the so-called SV (subjective value), which would be observed in our brains depending on the degree of "neuronal firing rate", which is generated when we perceive and evaluate said good or service to acquire it or not; and also that said utility or VS would be cardinal, not ordinal, and that it "learns", that is, it would improve day by day thanks to our "neuronal plasticity". That is, before this new empirical evidence, will continue maintaining the old neoclassical models?

CONCLUSIONS

Undoubtedly, the irruption of Neuroeconomics has come to stay. We are not talking about a passing fad, or a new fetish that economists are going to love for a while and then it will remain in the trunk of memories, as it has happened more than once. Advances in Cognitive Neuroscience are now available for all Social Sciences, and impacting and modifying theoretical traditions in fields as varied as Education, Marketing, Management, etc., aiming to cross transversally each and every one of the Social Sciences, which economics is hardly immune.

Now, to speak of a paradigm's shift in Kuhn's sense is unlikely, being in the sense of Lakatos much more pertinent, with changes in some topics, but not the whole economic field. In this way, throughout this work, we have been analyzing several of the lines of research where today Neuroeconomics most promises, in order to solve the main anomalies of the current neoclassical model. For example, Neuroeconomics is today trying to explain (among many others):

- the phenomenon of rigid wages falling;
- the anomalies in regards to intertemporal elections (excess of indebtedness at the family level, scarcity of savings for our retirement, etc.);
- strategic decisions;
- the influence of emotional aspects on individual demand functions;
- decisions in contexts of risk and uncertainty;

- and in general, specific aspects in each of the main fields of the traditional Economics, where the hypothesis of full rationality fails.

It happens that the irruption of Neuroeconomics and the Economics of Behavior, put in check the assumption of the neoclassical rationality (theory of the expected utility), underlying in almost all the base models of our science. We refer mainly to that cluster of ideas (long described throughout this book) that already at the end of the 19th century was entirely captured by the hand of Jevons, Walras, Marshall, etc.), and that were very well summarized by John Neville Keynes (the father of John Maynard Keynes, paradoxically a great critic of these postulates) of the following form:

- The correct methodological procedure of economics consists of starting from some fundamental facts about human nature. The point of beginning of the theories must be fundamentally observation, but introspection can also be useful, since it is considered as a source of obtaining ideas that can be described as empirical.
- The economic behavior that seeks self-interest dominates in reality the reasons for altruism and benevolence, so the economist must work knowing that man is selfish.
- The appropriate method for economics must end with the empirical observation relative to the fulfillment of the theory. However, the contrasts of the theories allow to determine their limits of application but not invalidate them: if a test, apparently, contradicts a theory, the researcher must be aware that this

133

result only shows that the test of that theory has been applied incorrectly.

This brief summary, made by J.N.Keynes (father J.M.Keynes, as we said) synthesizes almost perfectly the methodological position that prevailed among most economists in the 19th century, who we remember are the founders of the microeconomic theory, even today in force almost without fissures. This methodological posture will be further refined during the 20th century, starting with Lionel Robbins, and especially Milton Friedman, undoubtedly the most influential economist in the field of economic methodology during the 20th century.

With Friedman, in turn based on the remarkable Austrian philosopher Karl Popper, the theoretical shielding of traditional economic theory is finished, especially through the thesis of "irrelevance of the assumptions", where it is stated that it does not matter that the assumptions of a theory are realistic or not; the important thing is that the theory is able to predict accurately". Moreover, for Friedman it is verifiable that truly significant theories have premises that are clearly inadequate representations of reality and, in general, the more significant the theory, the less realistic their assumptions will be (as is the case of the imperfect hypothesis of rationality used in Economics for more than two centuries). For Friedman, the reason is simple, a hypothesis is important if explains a lot with little, that is, if it abstracts the important elements of the accessory.

And although at this point Friedman confuses something desirable (the simplicity of the model) with something undesirable (unrealistic premises), the penetration of this thesis in the standard methodology of economics is so great, that for current and future neuroeconomic models are accepted as a progress of science, is that they surpass in the matter of predictions to the traditional ones; otherwise, traditional modeling will continue in a certain way, despite many of its inconsistencies, since it is simpler than the neuroeconomic one.

In this way, there is a possibility that the current paradigm in economics does not change too much with this neuroeconomic boom; that will depend on that new models "predict better" than traditional ones. But why content ourselves with explaining the economic behavior of individuals based on false assumptions, in a time where neuroimaging and "transcranial magnetic stimulation", among other modern neurotechnics, allow us to specify with some approximation the real substrate of our decision making.

In the past, all those "discontented" with neoclassical rationality (J.M.Keynes, Simon and Hutchison, etc.) had no choice other than to appeal also to introspection to refute traditional thought; but now there are much more scientific instruments, such as those used by Neuroeconomics, impossible to ignore with the argument of "simplicity of the model".

Thus, throughout this book, we have mentioned several of these "dissatisfied" with the triumphant paradigm. For example, a "non-conformist ex ante" with the assumptions of the Neoclassical Economics is the very Adam Smith, the so-called "father of economics", who with his "Theory of Moral Sentiments" (written in the year 1756) expressed:

When I endeavor to examine my own behavior, when I endeavor to pronounce judgments on it, either to approve it or to condemn it, it is evident that in such cases it is as if I were divided into two different persons, and that I, the examiner and The judge embodies a man other than the other me, the person whose conduct is examined and judged.

The first is the spectator... The second is the agent, the person that I designate as myself, and from whose behavior I tried to form a feeling, as if it were a spectator's. The first is the judge, the second the person who is judged.

When we are about to act, the avidity of passion will rarely allow us to consider what we do with the dispassion of an intelligent person...

We also saw, in previous chapters, the critical thinking of another of the "discontented" with the current paradigm, Hutchison, who in 1938, with his work *The Significance and Basic Postulates of Economic Theory*, sharply criticizes neoclassical conventional wisdom:

*Simply to rely on **dogmatic assertions,** even when supported by phrases like "inner feelings of necessity" or "a priori facts", is to commit **scientific suicide**. It must really be explained in what precise way this "inner feeling of necessity," with which psychological method justifies its propositions, differs from the "inner felling of necessity" which **political fanatics** and the like always discover in support of their doctrines...*

We have seen that within Economics the "optimistic procedure of beginning with highly simplified "isolated" abstractions, in the hope of gradually making more realistic by removing the simplifying assumptions, is apt to come to a dead end, and that if one wants to get beyond a certain high level of abstraction one has to begin more or less from the beginning with extensive empirical research...

And also, throughout this work, we have analyzed the thinking of the contemporary Simon, who from a series of works that made him creditor to the Nobel Prize (not many years ago), gives an account of his questions to the principle of rationality in the decisions of businessmen, through his idea of "bounded rationality", where he states that, instead of optimizing in the way that neoclassical theory assumes, economic agents set a goal. When they achieve it, even if it is not optimal, they feel satisfied with it and do not seek to optimize. The men of flesh and bone have limited capacities to acquire knowledge and to make calculations, and to predict their behavior, from the theoretical point of view, it would be necessary the participation of psychologists and sociologists, in addition to the economists.

But in spite of its remarkable critical power, it is easy to see that both Adam Smith, Hutchison and Simon (the three just mentioned), as well as J.M. Keynes, among others of the critics profusely quoted throughout this work, ran into the great difficulty of not having scientific mechanisms of stem, superior to simple introspection, to refute neoclassical thought. However, now the question has changed, there are the neuroimages and the "trasncranean magnetic stimulation", among other available techniques, so modeling and debating with more theoretical support becomes easier. In fact, these advances are allowing Behavioral Economics, with the Nobel Kahneman at the head, to base much better a lot of research previously backed only on Psychology, but now also of Cognitive Neurosciences.

We believe that modeling in economics should involve considering the maximization of the affective and deliberative systems at the same time (either acting in a conflict or in a unitary form), but not only modeling the deliberative, as has been the tradition in Economics from the Neoclassical until now.

In short, the challenge for the coming decades is to create simplified neuroeconomic models that aim to consider the true neuropsychological aspects that underlie our decision making (something forgotten by economic science for more than two centuries), and with the objective that they can effectively be considered by the theoretical tradition in economics as useful, and in this way help to correct the many existing anomalies and theoretical inconsistencies. The challenge is great, but we believe not impossible, the tools are (still perfectible) and the anomalies of the traditional theory as well. Doing so, is almost a scientific obligation, whether or not the Friedman thesis is overcome.

BIBLIOGRAPHY

1. ADOLPHS R, TRANEL D, DAMASIO H, DAMASIO A. *Fear and the human amygdala.* Neuroscience, 1995; pág: 5879-91.

2. AHARON Itzhak, ETCOFF Nancy, ARIELY Dan, CHABRIS Chris F., O'CONNOR Ethan, and BREITER Hans C. 2001. "Beautiful Faces Have Variable Reward Value: fMRI and Behavioral Evidence." Neuron, 32(3): 537–51.

3. Akerlof, G.A. (1991), Procastination and obedience, American Economics Review, 8(2), 1-19.

4. ALEXANDER MP, BENSON DF, STUSS DT. Frontal lobes and language. Brain Lang 1989; 37: 656-91.

5. ANSLIE, G. (1992), Psicoeconmics, New York: Cambrigde University Press.

6. ANTONIETTI, Alessandro, Do Neurobiological Data Help Us to Understand Economic Decisions Better? Journal of Economics Methodology, Volume 17, Issue 2 June 2010, pages 207-218.

7. ARCHIBALD, G. C. (1959) "The State of Economic Science". *British Journal for the Philosophy of Science*, 10. Reimpreso en Marr, W.L. y Raj, B. (eds.), *How Economists Explain. A Reader in Methodology.* University Press of America, Lanham.

8. AUMANN, R. (2005) "War and Peace". Prize Lecture. http://nobelprize.org/nobel_prizes/economics/laureates/

9. AYDINONAT, N. EMRAH; Neuroeconomics: More Than Inspiration, Less Than Revolution; Journal of Economics Methodology, Volume 17, Issue 2 June 2010; pages 159-169.

10. BARBER, B. y T. ODEAN (2001). "Boys Will Be Boys: Gender, Overconfidence and Common Stock Investment", The Quarterly Journal of Economics.

11. BARON-COHEN S, LESLIE A M, FRITH U. *Does the autistic child have a 'Theory of Mind'?* Cognition 1985; 21: 37-46.

12. BARON-COHEN S. *Theory of Mind and autism: a review.*Special Issue of the International Review of Mental Retardation, 23, 169, (2001).

13. BERNHEIM B., "Neuroeconomics: A Sober (but hopeful) Appraisal"; Working Paper N° 13954; National Bureau of Economic Research; 2008.

14. BERRIDGE, Kent C. 1996. "Food Reward: Brain Substrates of Wanting and Liking." Neuroscience and Biobehavioral Reviews.

15. BHATT, Meghana and CAMERER, Colin; "Self-Referential Strategic Thinking and Equilibrium as States of Mind in Games: Evidence from fMRI." Games and Economic Behavior Journal.

16. BLAIR, R. James and CIPOLOTTI Lisa. 2000. "Impaired Social Response Reversal: A Case of 'Acquired Sociopathy.'" Brain, 123(6): 1122–41.

17. BLAUG, M. (1976) "Kuhn versus Lakatos o paradigmas versus programas de investigación en la historia de la economía pura". *Revista Española de Economía* 6, (primera época), enero-abril, 9-50.

18. BLAUG, M. (1992), *The Methodology of Economics.* (Segunda edición), Cambridge University Press, versión pdf.

19. BOEREEGeorge, Psicología General "Neurotransmisores", Universidad de Shippensburg.

20. BOLAND, *Critical Economic Methodology.* Routledge, Londres, 1997.

21. BRAIDOT, Nestor "Neuromarketing, Neuroeconomía y Negocios", Editorial Puerto Norte-Sur 2005.

22. BRAIDOT, Nestor, artículo en Revista "Entorno Económico", Mendoza, Argentina, febrero 2006.

23. BREITER, Hans C., AHARON Itzhak, KAHNEMAN Daniel, DALE Anders, and

SHIZGAL Peter. 2001. "Functional Imaging of Neural Responses to Expectancy and Experience of Monetary Gains and Losses", Revista Neuron.

24. CAIRNES, J.E. (1875) "The Character and Logical Method of Political Economy". MacMillan, Londres.

25. CALDWELL, B.J. (1994) *Beyond Positivism*. (Edición revisada), Routledge, Londres.

26. CAMERER C. F., Behavioral Game Theory Experiments in Strategic Interaction, Princeton University Press, 2003. http://press.princeton.edu/titles/7517.html

27. CAMERER C. F., THALER R.H., Anomalies: ultimatums, dictators and manners, Journal of Econ. Perspect.1995.

28. CAMERER, C. y LOEWENSTEIN, G. (2004). "Behavioral Economics: Past, Present, Future." Princeton: Princeton University Press.

29. CAMERER, C., LOEWENSTEIN, G. y PRELEC, D. (2005), "Neuroeconomics: How Neuroscience can inform Economics", *Journal of Economic Literature*. Vol. XLIII. No. 1.

30. CAMERER, C.; L. BABCOCK; G. LOEWENSTEIN y R. THALER (1997). "Labor Supply of New York City Cabdrivers: One Day at a Time"; The Quarterly Journal of Economics, 112(2), 407-441.

31. CARDONA HERRERO, Sergio; "Neuromanagement. Los Conocimientos sobre el Cerebro Aplicados al Mando en las Organizaciones" Editorial Almuzara, 2008.

32. CARSTENSEN, L.L., ISSACOWITZ, D M., CHARLES, S.T., 1999, *Taking time seriously: a theory of socioemotional selectivity*, American Psychologist 54, 165-181.

33. CARTER, Rita. El nuevo mapa del cerebro. Guía ilustrada de los descubrimientos más recientes para comprender el funcionamiento de la mente.

Editorial Integral. Ediciones de Librerías S.A. Barcelona, España, 1998.

34. CHABRIS, Christopher, LAIBSON, D. and SCHULDT, J., Decisiones Intertemporales, diciembre 2006, artículo publicado en Durlauf, S. y Blumen L, (Eds) (2007) The New Palgrave Dictionary of Economics (2nd edition), London: Palgrave Macmillan.

35. CHIC GARCÍA Genaro Neuroeconomía: Nuevas Orientaciones en los Estudios de Historia Económica. 2006.

36. CHORBAT, T. y McCabe, K. (2005). "Neuroeconomics and Rationality". *George Mason University School of Law". Working Paper Series. Paper 29.*

37. COHEN, J. (2005). "The Vulcanization of the Human Brain". *Journal of Economic Perspectives.* Vol 19. No. 4.

38. COOPER, R.W. y JOHN, A. (1988) "Coordinating Coordination Failures in Keynesian Models". *Quarterly Journal of Economics* 103, agosto, 441-463.

39. DAMASIO, Antonio "En busca de Spinoza. Neurobiología de la emoción y los sentimientos", Editorial Crítica, 2005.

40. DAMASIO, Antonio R. "Descartes' Error: Emotion, Reason, and the Human Brain", 1994, New York.

41. DE WAAL F. Good natured: the origins of right and wrong in humans and other animals. Cambridge: Harvard University Press; 1996.

42. DELGADO, Mauricio R., LEIGH E. NYSTROM, C. FISSELL, D. C. Noll, and Julie A. Fiez. 2000. "Tracking the Hemodynamic Responses to Reward and Punishment in the Striatum." Journal of Neurophysiology.

43. DELLA Vigna, Stefano and ULRIKE Malmendier. 2003, "Overestimating Self-Control: Evidence from the Health Club Industry," Berkeley Working Paper.

44. DENBURG, Natalie L., Psychophysiological anticipation of positive outcomes promotes advantageous decision-making in normal older persons, Elsevier. 2006.

45. DÍAZ ATIENZA Joaquín "Funciones Ejecutivas y Aprendizaje: I) Neuroanatomía y Evaluación" Extraído de: http://www.tdah-andalucia.es

46. ERK, Suzanne, SPITZER Manfred, WUNDERLICH Arthur P., GALLEY Lars, and WALTER Henrik. 2002. "Cultural Objects Modulate Reward Circuitry." Neuroreport.

47. FEHR Ernst, FISCHBACHER Urs and KOSFELD, Michael "Neuroeconomic Foundations of Trust and Social Preferences" 2005 Institute for the Study of Labor (IZA)

48. FISCHER, S. (1977) "Long-Term Contracts, Rational Expectations, and the Optimal Money Supply Rule". Journal of Political Economy 85, 1, 191-205.

49. FLETCHER, Paul C., HAPPE Francesca, FRITH Uta, BAKER S. C., DOLAN Ray J., FRACKOWIAK Richard S., and FRITH Chris D. 1995. "Other Minds in the Brain: A Functional Imaging Study of "Theory of Mind" in Story Comprehension." Cognition, 57(2): 109–28.

50. FLORES SÁNCHEZ, María de Lourdes "El aprendizaje acelerado".

51. FRIEDMAN, M. (1953); The Methodology of Positive Economics. En *Essays on Positive Economics*, University of Chicago Press, Chicago, 3-43. También disponible en: http://www.ppge.ufrgs.br/giacomo/arquivos/eco02 036/friedman-1966.pdf

52. FRIEDMAN, M. (1968) "The Role of Monetary Policy". *American Economic Review* 58, marzo, 1-17.

53. FRIEDMAN, M. (1990) *Teoría de los Precios*. Alianza Editorial, 2da edición.

54. FRITH, Uta. 2001. "Mind Blindness and the Brain in Autism." Neuron.

55. FRITH, Uta. 2001. "What Framework Should We Use for Understanding Developmental Disorders?" Developmental Neuropsychology, 20(2): 555–63.

56. FUMAGALLI, Roberto; The Disunity of Neuroeconomics: a Methodological Appraisal; Journal of Economics Methodology, Volume 17, Issue 2 June 2010; pages 119-131.

57. GLIMCHER P. W. "Decisions, uncertainty, and the brain: The science of neuroeconomics" 2003. Cambridge, MA: MIT Press.

58. GLIMCHER Paul W. and RUSTICHINI Aldo "Neuroeconomics: The Consilience of brain and Decision" 2004 Science Vol 306

59. GLIMCHER, P. (2003), *"Decisions, Uncertainty and the Brain. The Science of Neuroeconomics"*, Cambridge, Mass.: The MIT Press.

60. GLIMCHER, P.; CAMERER, C.; FEHR, E; POLDRACK, R.; *Neuroeconomics. Decision Making and the Brain. Editorial Elsevier, año 2009.*

61. GLIMCHER P., Choice: Towards a Standard Back Pocket Model, incluido en Neuroeconomics, Decision Making and the Brain, Elsevier, 2009.

62. GÓMEZ LÓPEZ, R., Evolución Científica y Metodológica de la Economía: Escuelas de Pensamiento, versión pdf, disponible en http://www.eumed.net/cursecon/libreria/rgl-evol/rgl-metod.pdf

63. GREENE JD, SOMMERVILLE RB, NYSTROM LE, DARLEY JM, COHEN JD. An fMRI investigation of emotional engagement in moral judgment. Science 2001.

64. GUL F. y PESENDORFER W., "The Case for Mindless Economics", Working Paper, Princeton University, 2005.

65. HAPPÉ F, BROWNELL H, WINNER E. Acquired 'Theory of Mind' impairments following stroke. Cognition 1999.

66. HAPPE, Francesca, EHLERS Stefan, FLETCHER Paul, FRITH Uta, JOHANSSON Maria, GILLBERG Christopher, DOLAN Ray, FRACKOWIAK Richard, and FRITH Chris. 1996. "'Theory of Mind' in the Brain: Evidence from a PET Scan Study of Asperger Syndrome." Neuroreport, 8(1): 197–201.

67. HARRISON, Glenn y ROSS, Don, "The Methodologies of Neuroeconomics", Journal of Economics Methodology, Volume 17, Issue 2 June 2010.

68. HSU, Ming; CAMERER, Colin y otros; 2005, "Ambiguity Aversion in the Brain: FMRI and Lesion Patient Evidence.", Caltech Working Paper.

69. HUANG Peter H. "Law and Human Flourishing: Happiness, Affective Neuroscience, and Paternalism".

70. HUME, D. (1980) [1748]. *"Investigaciones sobre el conocimiento humano"*. Madrid: Alianza Universidad.

71. HUTCHISON, T.W (1938) *The Significance and Basic Postulates of Economic Theory*. Edición de 1965: Augustus M. Kelley, Nueva York.

72. IUDICA, Valentina (1993), *Metodología y Epistemología de la Ciencia Económica*, Facultad de Ciencias Económicas, Universidad Nacional de Cuyo, Mendoza, Argentina.

73. JEVONS, W.S. (1871) *La Teoría de la Economía Política*, 1998, Editorial Pirámide

74. KAHNEMAN, D. (2003). "Maps of Bounded Rationality: Psychology for Behavioral Economics. American Economic Review. Vol 93. no. 5.

75. KAHNEMAN, D. and A. TVERSKY (1974). "Judgement Under Uncertainty: Heuristics and Biases", Science, Vol. 185, pág. 1124-1131.

76. KAHNEMAN, Daniel: "Maps of bounded rationality: psychology for behavioral economics", American Economic Review, 93, 5, 2003.

77. Kahneman D., *Remarks on Neuroeconomics,* incluido en *Neuroeconomics, Decision Making and the Brain,* Elsevier, 2009.

78. KEYNES, J.M. (1936); Teoría General del Empleo, el Interés y el Dinero, Fondo de Cultura Económica, 3ra edición 2001, disponible en la web en: http://www.listinet.com/bibliografia-comuna/Cdu332-38FB.pdf

79. KEYNES, J.N. (1890), *The Scope and Method of Political Economy.* 4ta Edición de 1915, University of Cambridge, disponible en: http://socserv.mcmaster.ca/econ/ugcm/3ll3/keynesj n/Scope.pdf

80. KIYOTAKI, N. (1988) "Multiple Expectational Equilibria under Monopolistic Competition". *Quarterly Journal of Economics* 102, noviembre, 695-714.

81. KNIGHT, F. (1940) "What is Truth in Economics?"; University of Chicago Press, 1956, p 151-178, disponible en Google Books, http://books.google.com.ar/books?id=rCyFf7vnH0U C&printsec=frontcover&dq=KNIGHT+Frank+What +is+Truth+in+Economics&hl=es&sa=X&ei=DrXvUtT CMIulsASKioC4Dw&ved=0CCoQ6AEwAA#v=one page&q=KNIGHT%20Frank%20What%20is%20Tru th%20in%20Economics&f=false.

82. KNUTSON, B., Rick S., WIMMER, G., PRELEC, D., LOEWENSTEIN G. (2006). Neural predictors of purchase. Neuron 53, 147–156

83. KNUTSON, Brian and PETERSON Richard. In Press. "Neurally Reconstructing Expected Utility." Games and Economic Behavior.

84. KNUTSON, Brian; WIMMER, G. ELLIOTT; KUHNEN, Camelia M.; WINKIELMAN, Piotr "Nucleus accumbens activation mediates the influence of reward cues on financial risk taking", 2008.

85. KNUTSON, LOEWENSTEIN y otros, Neural Predictors of Purchases, en Revista Neuron, enero de 2007.

86. KOENIGS, M., YOUNG, L., ADOLPH, R., TRANEL, D., CUSHMAN, F., HAUSER, M., DAMASIO, A. (2007) "Damage to the prefrontal cortex increases utilitarian moral judgemets". *Nature.* Vol 446.

87. KOOPMANS, T. (1957) *Three Essays on the State of Economic Science.* McGraw-Hill, Nueva York.

88. KUHN, T. (1977) *La estructura de las revoluciones científicas.* Fondo de Cultura Económica, México.

89. KUHNEN, C. y KNUTSON, B. (2005). "The Neural Basis of Financial Risk Taking". *Neuron.* Setiembre.

90. KUORIKOSKI, Jaakko y YLIKOSKI, Petri, Explanatory Relevance Across Disciplinary Boundaries: The Case of Neuroeconomics, Journal of Economics Methodology, Volume 17, Issue 2, June 2010, pages 219-228.

91. LAIBSON. D. (1997), Golden eggs and hyperbolic discounting. Quaterly Journal of Economics, 112, 443-477.

92. LAKATOS (1975) "La falsación y la metodología de los programas de investigación científica". En Lakatos, I. y Musgrave, A., (eds.) *La crítica y eldesarrollo del conocimiento*, Grijalbo, Barcelona, 203-343.

93. LAKATOS, I. (1971) "History of Science and Its Rational Reconstruction". En Cohen, R.S., Buck, C.R., Dordrecht-Holland, D. (eds.), *Boston Studies in Philosophy of Science* VIII.

94. LAKATOS, I. y MUSGRAVE, A. (eds.) (1975) La crítica y el desarrollo del conocimiento. Grijalbo, Barcelona.

95. LAMBRECHT, Anja and SKIERA Bernd. "Paying Too Much and Being Happy About It: Causes and Consequences of Tariff Choice–Bias." 2006. Johann Wolfgang Goethe-University Frankfurt.

96. LAVIN, Claudio, Emociones y decisión: marcadores somáticos, publicado en www.neuroeconomia.cl, año 2007.

97. LAZA, Sebastián, artículos públicados en blog ECONOMÍA APTA PARA TODO PÚBLICO: www.seblaza.blogspot.com.ar :

98. LESLIE A M.; Presence and representation: the origins of 'Theory of Mind'. Psychol Rev 1987; 94: 412-36.

99. LIEBERMAN Philip, Human Language and our reptilian Brain, Harvard Uniersity Press, Cambridge, 2002. http://books.google.com.ar/books?id= VyLdaH_Jw 0C

100. LIPSEY, R.G. (1974), Introducción a la Economía Positiva. Novena edición castellana: Vicens Vives, Barcelona. p. 15.

101. LIPSEY, R.G. (1991) Introducción a la Economía Positiva. Décimo segunda edición castellana: Vicens Vives, Barcelona.

102. LOEWENSTEIN, George F. 1994. "The Psychology of Curiosity: A Review and Reinterpretation." Psychological Bulletin.

103. LOEWENSTEIN, George F., WEBER Robert, FLORY Janine, MANUCK Stephen and MULDOON Matthew. 2001. "Dimensions of Time Discounting." Conference on Survey Research on Household Expectations and Preferences.

104. LOEWESTEIN, G. and PRELEC, D. (1992) Anomalies in intertemporal choice: evidence and an interpretation. Quaterly Journal of Econnomics, 107, 573-957.

105. LOEWESTEIN y O'DONOGHUE "Animal Spirits: Affective and Deliberative Processes in Economic Behavior" 2005.

106. LÓPEZ, Ernesto, "Todos tenemos nuestro cuarto de hora: Economía Conductual, Neuroeconomía y sus

implicancias para la protección al consumidor", mimeo, año 2005, INDECOPI.

107. MACHLUP, F. (1978) "Operationalism and Pure Theory in Economics". En *Methodology of Economics and Other Social Sciences*. Academic Press, Nueva York.

108. MÄKI, Uskali; When Economics Meets Neuroscience: Hype *And* Hope; Journal of Economics Methodology, Volume **17**, Issue **2** June 2010; pages 107-111.

109. MANKIW, N.G. (1990) "A Quick Refresher Course in Macroeconomics". *Journal of Economic Literature* 28, diciembre, 1645-1660.

110. MARCHIONNI, Caterina y VROMEN, Jack; 'Neuroeconomics: Hype or Hope?', Journal of Economics Methodology, Volume 17, Issue 2 June 2010, pages 103-106.

111. MAS-COLELL, A., WHINSTON, M. y GREEN, J. (1995). "Microeconomic Theory". Oxford: Oxford University Press.

112. Mc LURE, Samuel, LAIBSON David, LOEWENSTEIN George and COHENJonathan, Separate neural system value inmediate and delayed monetary rewards, Science, octubre de 2004.

113. McCABE, K., HOUSER, D., RYAN, L., SMITH, V. y TROUARD, T. (2001) "A functional imaging study of cooperation in two person reciprocal exchange". *Proceedings of the National Academy of Sciences of the United States of America".* www.pnas.org/cgi/doi/10.1073/pnas211415698.

114. McCLURE, Samuel M., LAIBSON David, LOEWENSTEIN George, and COHEN Jonathan D. "Separate Neural Systems Value Immediate and Delayed Monetary Rewards" 2004.

115. MELITZ, J. (1965) "Friedman and Machlup on the Significance of Testing Economic Assumptions". *Journal of Political Economy* 73, 37-60.

116. MENDOZA Lara Elvira y LÓPEZ HERRERO Paz Consideraciones sobre el desarrollo de la Teoría de la Mente (TOM) y del Lenguaje. Dpto. Personalidad, Evaluación y Tratamiento Psicológico. Universidad de Granada

117. MILL, J.S., *Principles of Political Economy* London: Longmans, Green and Co., ed. William J. Ashley, (1909, 7ma edición), disponible en: http://www.gutenberg.org/files/30107/30107-pdf.pdf

118. MILL, J.S. (1967) *Collected Works, Essays on Economic and Society*. J.M. Robson (edit). University of Toronto Press, Toronto, p. 323; también disponible en: http://files.libertyfund.org/files/244/Mill_0223-04_EBk_v7.0.pdf

119. Mobbs Dean, Greicius Michael D., Eiman Abdel-Azim, Vinod Menon, and Allan L. Reiss. 2003. "Humor Modulates the Mesolimbic Reward Centers." Neuron.

120. MOLL J, ESLINGER P, OLIVEIRA-SOUZA R. Frontopolar and anterior temporal cortex activation in a moral judgment task: preliminary functional MRI results in normal subjects. Arq Neuropsiquiatr 2001.

121. MONTAGUE, P. Read and Gregory S. BERNS. 2002. "Neural Economics and the Biological Substrates of Valuation." Neuron, 36(2): 265–84.

122. MORA TERUEL Francisco Neurocultura: todo está en el cerebro.2007 www.abc.es

123. MORALES, Raúl, Una nueva ciencia, la Neuroeconomía, estudia las decisiones económicas humanas, publicado en: http://www.secyt.unc.edu.ar/Temas/Boletin/articulos/BoletinV1N10_neuroeconomia.html, agosto de 2003, originalmente publicado en Tendencias Científicas (21/06/03) http://www.tendencias21.net

124. MOTTERLINI Mateo "Economía emocional, En qué nos gastamos el dinero y por qué" Editorial Paidós Ibérica.2008

125. MULLAINATHAN, S. y R. THALER (2000). "Behavioral Economics", Working Paper 7948. National Bureau of Economic Research.

126. MUTH, J. (1961) "Rational Expectations and the Theory of Price Movements". Econométrica 39, julio, 315-334.

127. NAGATSU, Michiru, Function and Mechanism: The Metaphysics of Neuroeconomics, Journal of Economics Methodology, Volume 17, Issue 2 June 2010, pages 197-205.

128. NAVARRO, Alfredo Martín, De SCHANT, Fermín y MARTÍN, Jorge Marcelo, Neuroeconomía y Metodología: Algunas Reflexiones Iniciales, año 2007, mimeo.

129. NAVARRO, Alfredo, Neuroeconomía y Teoría de Juegos. Implicancias Metodológicas, mimeo (2007).

130. OLDS, James and MILNER Peter. 1954. "Positive Reinforcement Produced by Electrical Stimulation of Septal Area and Other Regions of Rat Brain." Journal of Comparative and Physiological Psychology

131. PATINKIN, D. (1959), *Dinero, Interés y Precios,* Editorial Aguilar.

132. PEYROLÓN, Pablo, Neuroeconomía o la Economía del Prozac, Profesor Asociado Universitat Pompeu Fabra y ESCI, publicado en http://www.eumed.net/ce/pp-neuro.htm.

133. PHELPS, E. (1967) "Phillips Curves, Expectations of Inflation and Optimal Unemployment over Time". *Economica* 34, agosto, 254-281.

134. POPPER (1991), *Discurso de Investidura como Doctor Honoris Causa,* Universidad Complutense, Madrid.

135. POPPER, K. (1962) La lógica de la investigación científica, Tecnos, Madrid, disponible en: http://ifdc6m.juj.infd.edu.ar/aula/archivos/repositor

io//0/103/Karl R. Popper -
La Logica de la Investigacion Cientifica.pdf

136. PRELEC, Drazen and Duncan Simester. 2001. "Always Leave Home without It." Marketing Letters.

137. PRELEC, Drazen and George F. LOEWENSTEIN. 1998. "The Red and the Black: Mental Accounting of Savings and Debt." Marketing Science.

138. RABIN M., A perspective on psychology and economics, Eur. Econ. Rev. 46 (2002).

139. RAMÓN, José María (2004), *La Epistemología de Khun, Lakatos y Feyerabend: un análisis comparado*, Universidad Nacional de la Patagonia, pág 53-62, disponible en: http://josemramon.com.ar/wp-content/uploads/Ram%C3%B3n-Jos%C3%A9-Mar%C3%ADa-La-epistemolog%C3%ADa-de-khun-Lakatos-y-Feyerabend.pdf

140. RILLING, J., GUTMAN, D., ZEH, T., PAGNONI, G., BERNS, G., y KILTS, C. (2002). "A Neural Basis for Social Cooperation". *Neuron*. Vol 35. No. 2.

141. ROBBINS (1971) *Autobiography of an Economist*. Macmillan, Londres.

142. ROBBINS, L. (1932) *An essay on the Nature and Significance of Economic Science*, MacMillan, Londres, disponible en: http://mises.org/books/robbinsessay2.pdf y reeditado por The Mises Institute, Alabama, 2007, disponible en Google Books: http://books.google.com.ar/books?id=nySoIkOgWQ4C&printsec=frontcover&dq=lionel+robbins&hl=es&sa=X&ei=Y7DvUr7-AcipsQTolIDgBw&sqi=2&ved=0CC0Q6AEwAA#v=onepage&q=lionel%20robbins&f=false

143. ROBBINS, L. (1934) "Remarks on the Relationship between Economics and Psychology". The Manchester School of Economics and Social Science.

144. ROTWEIN, E. (1959) "On the Methodology of Positive Economics, *QuarterlyJournal of Economics* 73, 554-575.

145. ROWE, Andrea D., BULLOCK Peter R., POLKEY Charles E., and MORRIS Robin G.. 2001. "'Theory of Mind' Impairments and their Relationship to Executive Functioning Following Frontal Lobe Excisions." Brain.

146. SAMUELSON (1963) Problems of Methodology: Discussion. *American* Economic Review Papers and Proceedings 53, 2, 231-236.

147. SAMUELSON, P.A. (1948) *Foundations of Economic Analysis*. Harvard University Press, Cambridge, Mass.

148. SÁNCHEZ-ROBLES, Blanca, La Economía. Concepto y Método, disponible en: personales.unican.es/sanchezb/web/La%20economi a.pdf sitio de la Universidad de Cantabria, España.

149. SANFEY, Alan G., James K. RILLING, Jessica A. AARONSON, Leigh E. NYSTROM, and Jonathan D. COHEN. 2003. "The Neural Basis of Economic Decision-Making in the Ultimatum Game." Science,

150. SANFEY, Rilling, COHEN y otros, The Neural Basis of Economic Decision Making in the Ultimatum Game, en Revista Science, junio de 2003.

151. SANFEY, Rilling, COHEN y otros, The neural correlates of Theory of Mind within interpersonal interactions, en Revista Science Direct, año 2004.

152. SARGENT, T. y WALLACE, N. (1975) "'Rational Expectations', the Optimal Monetary Instrument, and the Optimal Money Supply Rule". *Journal of PoliticalEconomy* 83, abril, 241-254.

153. SAXE, Rebecca and KANWISHER Nancy. 2003. "People Thinking about Thinking People: The Role of the Temporo–Parietal Junction in 'Theory of Mind.'" Neuroimage, 19(4): 1835–42.

154. SCHULTZ, Wolfram. 2002. "Getting Formal with Dopamine and Reward." Neuron.

155. SCHUMPETER, J. (1971) *Historia del Análisis Económico*. Fondo de Cultura Económica, México.

156. SCHWARTZ, P. (1997) "Invitación a la economía". En Febrero, R. (edit.) *Qué es la Economía*, Pirámide, Madrid, 65-100.

157. SEN, A. (1987). "Rational Behavior"; Macmillan Press Limited.

158. SENIOR, N. (1827) Introductory Lecture on Political Economy. En Selected Writings on Economics. A Volume of Pamphlets 1827-1852. Kelley, Nueva York, disponible en: http://mises.org/books/selected_writings_senior.pdf

159. SENIOR, N. (1836) *Outline of the Science of Political Economy*, edición de 1951: Kelley, Nueva York, disponible en http://digamo.free.fr/senior36.pdf

160. SIEGAL M, CARRINGTON J, RADEL M. Theory of Mind and pragmatic understanding following right hemisphere damage. 1996. Brain and Language.

161. SIMON, H. (1997). *"An Empirically Based Microeconomics"*. Raffaelle Mattioli Foundation. Cambridge: Cambridge University Press, version pdf.

162. SINGER Tania and FEHR Ernst "Neuroeconomics of Mind Reading and Empathy " 2005 University of Zurich and IZA

163. SINGER, Tania, KIEBEL, Stefan y otros, 2004, "Brain Responses to the Acquired Moral Status of Faces", Neuron, 41 (4): 653-62.

164. SINGER, Tania; KIEBEL, STEFAN J.; WINSTON, Joel S.; DOLAN, Ray J. and FRITH, Christopher D. "Brain responses to the acquired moral status of faces". Neuron, February 2004a

165. SMITH, A. (1941) [1759]. "Teoría de los Sentimientos Morales". México: Fondo de Cultura Económica.

166. SPERRY, GAZZANIGA y BOGEN "Interhemispheric relationships, the neocortical

commisures: syndromes of hemisphere disconnection", en Handbook of Clinical Neurology, 1969, vol 4, pág 273-290, Amsterdam, North Holland.

167. STROZT, R. H. (1956) Myopia and inconsistency in dynamic utility maximization, Review of Economics Studies, 23, 165-180.

168. THALER, R. (1985). "Mental Accounting and Consumer Choice", Marketing Science, Vol. 4, pág. 199-214.

169. TIRAPU-USTÁRROZ J., PÉREZ-SAYES G., EREKATXO-BILBAO M., PELEGRÍN-VALERO C. ¿Qué es la teoría de la mente? 2007. Revista de Neurología.

170. TRAIN, Kenneth E. Optimal Regulation: The Economic Theory of Natural Monopoly. 1991 Cambridge: MIT Press.

171. TRAIN, Kenneth E., Daniel L. McFADDEN, and Moshe BEN-AKIVA. 1987. "The Demand for Local Telephone Service: A Fully Discrete Model of Residential Calling Patterns and Service Choices." Rand Journal of Economics.

172. TURNER, Jonathan H; On the origins of Human Emotions, Stanford University Press, 2000.

173. VARIAN Hal R; "Microeconomía intermedia"; 4ta Edición, Editorial Antoni Bosch 1996.

174. VARIAN, Hal R; "Microeconomic Analysis"; 1978.

175. VERCOE, Moana y ZAK, Paul; Inductive Modeling Using Causal Studies in Neuroeconomics: Brains on Drugs; Journal of Economics Methodology, Volume 17, Issue 2; June 2010.

176. VerLEE, W.L.; Aprender con todo el cerebro; Ediciones Martínez Roca; 1986.

177. WINNER E, BROWNELL H, HAPPÉ F, BLUM A, PINCUS D.; *Distinguishing lies from jokes: Theory of Mind deficits and discourse interpretation in right hemisphere brain-damaged patients.* Brain Lang 1998; 62: 89-106.

178. VROMEN, Jack; Where economics and neuroscience might meet; Journal of Economics Methodology, Volume 17, Issue 2 June 2010; pages 171-183.

179. ZAK Paul J. "Neuroeconomics" 2004 The Royal Society

180. ZAK y FAKHAR, Neuroactive hormones and interpersonal trust: international evidence, Elsevier, año 2006.

181. ZALTMAN Gerald, Cómo Piensan los Consumidores, Ediciones Urano, S.A. Editorial Empresa Activa. 2004.

NOTAS

[1]Kahneman, (2003): "Maps of Bounded Rationality: Psychology for Behavioral Economics", American Economic Review, 93.

[2]Zak, (2004), "Neuroeconomics". Paul Zak is one of the most renowned neuroeconomists today, specialized in the relationship between oxitocyn and trust.

[3]Glimcher Paul W. and Rustichini Aldo "Neuroeconomics: The Consilience of Brain and Decision", (2004), Science Vol 306.

4 Neuroeconomist of the George Washington University.

5 Extracted from Tom Wolfe, Hooking Up (2000).

[6]Schumpeter, J. (1971) History of Economic Analysis. Fondo de Cultura Económica, Mexico, p. 167.

[7]Braidot, Nestor; article in "Entorno Económico" Magazine, Mendoza, Argentine, February 2006.

8 C.F. Camerer, G. Loewenstein, D. Prelec, "Neuroeconomics: How Neuroscience can Inform Economics", Working Paper, UCLA Department of Economics, (2003).

9 Hume, D. (1980) [1748]. "Investigations on human knowledge". Madrid: University Alliance.

10 Smith, A. (1941) [1759]. "Theory of moral feelings". Mexico: Fondo de Cultura Económica.

11 Simon, H. (1997). "An Empirically Based Microeconomics". Raffaelle Mattioli Foundation. Cambridge: Cambridge University Press, pdf version.

[12]MILL, J.S. (1967) Collected Works, Essays on Economic and Society. J.M. Robson (edit). University of Toronto Press, Toronto, p. 323

13 KEYNES, J.M., General Theory of Employment, Interest and Money, Fondo de Cultura Económica, 3rd edition 2001, available on the web at: http://www.listinet.com/bibliografia-comuna/Cdu332-38FB.pdf

14 PATINKIN, D. (1959), Money, Interest and Prices, Editorial Aguilar.

15 Simon, H. (1997), "An Empirically Based Macroeconomics," Raffaelle Mattioli Foundation. Cambridge: Cambridge University Press, pdf version.

16 See Kahneman, D. (2003). "Maps of Bounded Rationality: Psychology for Behavioral Economics. American Economic Review. Vol 93. N° 5, where you summarize the results of your investigations.

[17]Camerer, C. and Loewenstein, G. (2004). "Behavioral Economics: Past, Present, Future" in Camerer C. and Lowenstein G. (ed.) "Advances in Behavioral Economics", Princeton: Princeton University Press.

[18] López, Ernesto, "We all have our quarter of an hour: Behavioral Economics, Neuroeconomics and its implications for consumer protection", mimeo, year 2005, INDECOPI, p. 119-120.

[19] MULLAINATHAN, S. and R. THALER (2000). "Behavioral Economics", Working Paper 7948. National Bureau of Economic Research.

[20] KAHNEMAN, D. and A. TVERSKY (1974). "Judgment Under Uncertainty: Heuristics and Biases", Science, Vol. 185, p. 1124-1131.

[21] THALER, R. (1985). "Mental Accounting and Consumer Choice", Marketing Science, Vol. 4, p. 199-214.

[22] CAMERER, C.; L. BABCOCK; G. LOEWENSTEIN and R. THALER (1997). "Labor supply of New York City cabdrivers: One day at a time", The Quarterly Journal of Economics, 112 (2), 407-441.

[23] Loewenstein, G. and O'Donoghue, T. (2004). "Animal Spirits: Affective and Deliberative Influences on Economic Behavior". Working Paper.

[24] http://discovermagazine.com/2014/april/14-the-second-coming-of-sigmund-freud

[25] Camerer, Loewestein and Prelec (2005); Neuroeconomics: How Neuroscience Can Inform Economics; Journal of Economics Literature, Vol. XLIII, N°1.

[26] López, Ernesto, "We all have our quarter of an hour: Behavioral Economics, Neuroeconomics and its implications for consumer protection", mimeo, year 2005, INDECOPI, p. 114-116.

[27] BARBER, B. and T. ODEAN (2001). "Boys Will Be Boys: Gender, Overconfidence and Common Stock Investment", The Quarterly Journal of Economics.

[28] Kuhnen, C. and Knutson, B. (2005), "The Neural Basis of Financial Risk Taking", Neuron. September.

[29] Hsu, Ming; Camerer, Colin and others; 2005, "Ambiguity Aversion in the Brain: FMRI and Lesion Patient Evidence.", Caltech Working Paper.

[30] Camerer, C., Loewenstein, G. and Prelec, D. (2005), "Neuroeconomics: How Neuroscience can inform Economics", Journal of Economic Literature. Vol. XLIII. N° 1.

[31] McCabe, K., Houser, D., Ryan, L., Smith, V. and Trouard, T. (2001) "A functional imaging study of cooperation in two person reciprocal exchange". Proceedings of the National Academy of Sciences of the United States of America, www.pnas.org/cgi/doi/10.1073/pnas211415698

[32] Singer, Tania, Kiebel, Stefan and others, 2004, "Brain Responses to the Acquired Moral Status of Faces", Neuron, 41 (4): 653-62.

[33] Alfredo Navarro, Neuroeconomics and Game Theory. Methodological Implications, mimeo (2007). The author is a Full Member of the National Academy of Economic Sciences.

[34] Glimcher, P. (2003), Decisions, Uncertainty and the Brain. The Science of Neuroeconomics, Cambridge, Mass.: The MIT Press.

[35] Glimcher (2003), "Decisions, Uncertainty and the Brain. The Science of Neuroeconomics", Cambridge, Massachussets: The MIT Press, p. 321.

[36] Sanfey, Rilling, Cohen and others, The Neural Correlates of Theory of Mind within Interpersonal Interactions, in Science Direct Magazine, 2004.

[37] Sanfey et al, The Neural Basis of Economic Decision Making in the Ultimatum Game, in Science Magazine, June 2003.

[38] Sanfey, Rilling, Cohen and others, The Neural Basis of Economic Decision Making in the Ultimatum Game, in Science Magazine, June 2003.

[39] Zak and Fakhar, Neuroactive Hormones and Interpersonal Trust: International Evidence, Elsevier, 2006.

[40] Loewenstein, G. and O'Donoghue, T. (2004). "Animal Spirits: Affective and Deliberative Influences on Economic Behavior". Working Paper.

[41] Cohen, J. (2005), "The Vulcanization of the Human Brain", Journal of Economic Perspectives, Vol 19. No. 4.

[42] Koenigs, M., Young, L., Adolph, R., Tranel, D., Cushman, F., Hauser, M., Damasio, A. (2007) "Damage to the prefrontal cortex increases utilitarian moral judgemets". Nature, Vol 446.

[43] Denburg, Natalie L., Psychophysiological anticipation of positive outcomes promotes advantageous decision-making in normal older persons, Elsevier, 2006.

[44] Carstensen, L.L., Issacowitz, D.M., Charles, S.T., 1999, Taking time seriously: a theory of socioemotional selectivity, American Psychologist 54, 165-181.

[45] Anslie, G. (1992), Psychoeconomics, New York: Cambrigde University Press.

[46] Loewestein G. and Prelec D. (1992) Anomalies in intertemporal choice: evidence and an interpretation. Quaterly Journal of Econnomics, 107, 573-957.

[47] Laibson. D. (1997), Golden eggs and hyperbolic discounting. Quaterly Journal of Economics, 112, 443-477.

[48] Mc Lure, Samuel, David Laibson, George Loewenstein and Jonathan Cohen, Separate neural system value immediate and delayed monetary rewards, Science, October 2004

[49] Aharon, Itzhak, Nancy Etcoff, Dan Ariely, Chris F. Chabris, Ethan O'Connor, and Hans C. Breiter. 2001. "Beautiful Faces Have a Variable Reward Value: fMRI and Behavioral Evidence." Neuron, 32 (3): 537-51. Mobbs, Dean, Michael D. Greicius, Eiman Abdel-Azim, Vinod Menon, and Allan L. Reiss. 2003. "Humor Modulates the Mesolimbic Reward Centers." Neuron. Erk, Suzanne, Manfred Spitzer, Arthur P. Wunderlich, Lars Galley, and Henrik Walter. 2002. "Cultural Objects Modulate Reverse Circuitry." Neuroreport.

Schultz, Wolfram. 2002. "Getting Formal with Dopamine and Reward." Neuron. Breiter, Hans C., I. Aharon, Daniel Kahneman, A. Dale, and Peter Shizgal. 2001. "Functional Imaging of Neural Responses to Expectancy and Experience of Monetary Gains and Losses." Neuron. Knutson, Brian and Richard Peterson. In Press. "Neurally Reconstructing Expected Utility." Games and Economic Behavior. Delgado, Mauricio R., Leigh E. Nystrom, C. Fissell, D. C. Noll, and Julie A. Fiez. 2000. "Tracking the Hemodynamic Responses to Reward and Punishment in the Striatum." Journal of Neurophysiology.

[50] Camerer, C., Loewenstein, G. and Prelec, D. (2005), "Neuroeconomics: How Neuroscience can Inform Economics", Journal of Economic Literature. Vol. XLIII. No. 1

[51] Glimcher P., Choice: Towards to Standard Back Pocket Model, included in Neuroeconomics, Decision Making and the Brain, Elsevier, 2009.

[52] Kahneman D., Remarks on Neuroeconomics, included in Neuroeconomics, Decision Making and the Brain, Elsevier, 2009.

www.ingramcontent.com/pod-product-compliance
Lightning Source LLC
Chambersburg PA
CBHW021414210526
45463CB00001B/362